Decolonising Mathematics Education

Anti-colonial Educational Perspectives for Transformative Change

VOLUME 14

The titles published in this series are listed at *brill.com/acep*

Decolonising Mathematics Education

Transforming Pedagogy alongside First Peoples of Remote Australia

By

Nicole Boyd

BRILL

LEIDEN | BOSTON

Cover illustration: Jedda Purvis (blue beanie), Judy Purvis (pink beanie) and Louisa Long (black beanie) responding to a student led inquiry. Photograph by Tim Ware.

All chapters in this book have undergone peer review.

The Library of Congress Cataloging-in-Publication Data is available online at https://catalog.loc.gov

Typeface for the Latin, Greek, and Cyrillic scripts: “Brill”. See and download: brill.com/brill-typeface.

ISSN 2542-9280
ISBN 978-90-04-68273-3 (paperback)
ISBN 978-90-04-68274-0 (hardback)
ISBN 978-90-04-68275-7 (e-book)

This book is printed on acid-free paper and produced in a sustainable manner.

To my father, Max Boyd, whose recent passing taught me how important it is to never change who you are

Contents

Acknowledgements IX
List of Figures and Tables XII
List of Abbreviations XVI
About the Author XVII

1 Introduction 1
1 Apmer Akely-Akely 4
2 Re-imagining Curriculum and Pedagogy 10

2 Validating Indigenous Pedagogies in Mathematics Education 29
1 Are First Nations Identities and Cultural Intelligences Invoked in Mathematics Learning? 29
2 A Framework for Validating Indigenous Pedagogy 33
3 Cultural Respect Framework 47
4 Critical Self-Reflection 50

3 Embedding Living Stories in Mathematics Education 63
1 Utopia Community 63
2 My Positionality – An Insider/Outsider 74
3 Why Was Intervention in My Pedagogic Practice Necessary? 77

4 Preferencing Utopia Eastern Anmatyerr and Alyawarr Voice in Pedagogy 89
1 Creating the Space and Place for Merging Epistemologies in Measurement Curriculum 91
2 Decolonising Mathematics Education 95

5 The Goompi Model 110
1 Goompi Iterative Pedagogical Cycles in Four Phases 110
2 Goompi Pedagogical Cycles: Akatyerr Quadrat Study Iterations 114

6 Narrative: The Experience and Story of Decolonising Mathematics Education 132
1 Narrative: The Experience of Change 132
2 Understanding the Experiences of Change 136

7 **Theory of Social Change: Decolonising Mathematics Education** 147
1 Major Findings 147
2 Solutions to Problematic Pedagogic Practices 153
3 The Contribution of the Theoretical Frameworks to the Study 158
4 The Contribution of the Thematic Analysis 161
5 Limitations of the Study 162
6 Recommendations for Mathematics Education 163
7 Directions for Further Research 164

8 **Future Generations** 170
1 Anti-Colonial Educational Perspectives for Transformative Change 170
2 Lessons for Future Teachers 174
3 Mathematisation of Wild Akatyerr Harvest 177
4 Future Questions 184
5 Summarising Thoughts 185

Glossary 191
Index 192

Acknowledgements

This book arose after I completed my Master of Philosophy thesis, *Two-Way Dialogue on Akatyerr (Desert Raisin) in a Female Indigenous Middle Years Class: Towards Cultural Inclusivity in Mathematics Curriculum and Pedagogic Practice*, at QUT in 2021. The thesis enabled me to critically reflect on my role as a high school (Years 8–9) mathematics teacher in a remote Australian Indigenous community. I was particularly interested in creating opportunities for students to represent mathematical understanding beyond the familiar multimodal forms: language, numeric, and visual (Payne & Rathmell, 1977). For this reason, I sought pedagogical practices that permit students to move flexibly between language, symbols, models, and real-world understandings of mathematics (YuMi Deadly Centre, 2014). My interest in using the real world of Indigenous students' lives as an entry point into the mathematics curriculum developed since I recognised that students could explore and see themselves, their identities, and their learning capabilities reflected in this curriculum. I am grateful for the opportunity provided by Koninklijke Brill Publishers to relate the stories contained within this book to the broader international audience.

I acknowledge and thank the late Emeritus Professor Tom Cooper for his encouragement, ideas, and support to pursue a higher degree research project when I first approached him.

I am eternally grateful to my thesis supervisor, Associate Professor Bronwyn Ewing. Thank you, Bron, for your relentless enthusiasm and support for my project ideas, meticulous proofreading of thesis drafts, patience, and good humour during the many setbacks to the progress of the study. I am grateful also to my associate supervisor, Dr Chrystal Whiteford. Thank you for your interest in this project and your guidance throughout thesis revisions.

I am deeply inspired by and grateful to Professor Chris Matthews for his advice, guidance, and encouragement with adapting and using his Goompi pedagogical cycle. I sincerely thank you, Chris, for all your encouragement and am grateful that you have provided opportunities to share this work through ATSIMA presentations. I am extremely honoured that you agreed to read the manuscript.

I am deeply grateful to bush food business pioneer Rayleen Brown for sharing with me her passion and support for intergenerational transfer of akatyerr knowledge. I thank you, Rayleen, for your ongoing support and enthusiasm for the establishment of a new Utopia Alyawarr ranger group to protect this valuable cultural resource.

I am humbled by the work and encouragement of Dr David Moore for his role in this study as a critical friend, a trustworthy and knowledgeable linguist who is a committed agent of change through his translation work. I sincerely thank linguists Dr David Moore, Mrs Susan Moore, Associate Professor Myf Turpin, and Dr Margaret Carew for your support and assistance with preparation for audio recording of participants' yarning circle dialogue.

I am sincerely grateful to Dr Terri Janke, solicitor director; Charisma Cubillo and Laura Curtis, solicitors; and Harry Loader, paralegal, of Terri Janke and Company. Thank you all for your contributions to writing the Indigenous cultural and intellectual property notices and providing advice and encouragement for the treatment of Indigenous knowledge shared in the study. I am deeply appreciative of the work provided when negotiating the publishing contract, the informed consent forms, and the image release deeds.

Thank you to Charlotte Cottier (accredited editor with the Institute of Professional Editors) for patiently waiting for this story to come along and being available to edit the thesis document. I am also deeply appreciative of your careful reading and professionalism when editing this manuscript.

Thank you to Stephen Nimmo (NTDOE, School Principal) and Suzanne Stone (NTDOE, Assistant Principal) for your support particularly with navigating and permitting changes to the high school timetable to allow the yarning circle research to occur. Thank you to Tim Ware (NTDOE), Cari Bertram (NTDOE), Paul Evans (NTDOE/CLC), Dr Fiona Walsh, Genevieve O'Loughlin, and Dave Albrecht for your permission to use your photographs. Thank you to Jimmy Navie from Wanta Aboriginal Corporation for your operational and collegial support throughout the research. Thank you to Lara Hynes for your good humour and skill with graphic design elements. I am appreciative of the lifelong friendship, support, and guidance with the construction and use of quadrat frames from Clorinda Schofield. Thank you to Paul Evans, my husband, for your love, belief in me, and continued support throughout the research.

Paul and I are appreciative of all our dear friends in Melbourne, especially Steve Parker, Bram Mason, Cheryl Taylor, and Clorinda Schofield, who listened eagerly while we shared countless stories. Finally, to our family – we thank you for letting us go to pursue our passion and seeking to understand our vision. We are eternally grateful for your unrelenting love and support.

Most importantly, thank you to the Utopia Eastern Anmatyerr and Alyawarr people, who welcomed Paul and me into their community and made us feel part of their family. I am deeply inspired by working alongside Sharmic. Her willingness to share cultural knowledge and continuously navigate between both worlds to develop students mathematical understanding is truly inspiring. Thank you also to the community members involved at the school site of research: and the Year 8/9 girls. Thank you especially to the two elders at the akatyerr research site, Jedda Purvis and Judy Purvis. Thank you also to the Utopia Alyawarr rangers at the akatyerr research site.

I am deeply grateful for the many conversations before, during, and after yarning circle consultations with assistant teachers.

Finally, thank you to the four students who agreed to be a part of the akatyerr research bush trip and to their families. I am deeply appreciative of your willingness to try new learning opportunities. I am immensely proud of the ease with which you navigate both worlds as you continue your learning.

Figures and Tables

Figures

1 Ngkwarl (sugarbag) is gently coaxed out of the bloodwood tree with a probing stick. 6
2 Wild akatyerr fruit (*Solanum centrale*). 8
3 Bucketful of dried akatyerr. 8
4 Praxis in Indigenous mathematics education. 10
5 Utopia Alyawarr ranger (fifth from left, wearing a yellow shirt) sharing stories about caring for his country. 14
6 Constructing new measurement knowledge in an intercultural space (adapted from Northern Territory Department of Education, 2017b, pp. 15–21). 16
7 Example question to demonstrate disciplinary thinking of area of the region in green. 31
8 Quadrat survey design to solve problems involving areas of common shapes. Modelling a practical situation – akatyerr vegetation cover is surveyed along a 50-metre transect line to construct new mathematics knowledge. 32
9 Principles of Indigenous research. The three principles – voice, political integrity, and resistance – are bound by a framework of cultural respect. 33
10 The human dimension to mathematics learning. This image depicts the interactions between community members, the teacher, and students in a familiar social context, the yarning circle. 41
11 Jimmy Navie (right) from Wanta Aboriginal Corporation helps two students construct new mathematics knowledge about sustainable wild akatyerr harvest using the quadrat survey. 42
12 Knowledge concepts interrelated to akatyerr (adapted from Dobson et al., 2008), Merne Altyerre-ipenhe Reference Group et al., 2011, and Turner, 2005. 49
13 Proportional reasoning – solving the percentage of plant cover in the quadrat frame. 53
14 Akatyerr I noticed growing in disturbed roadside sandy red soils near where I lived. The yellow fruit are ripe, the green fruit are unripe. 64
15 Students prepare bush medicine plants for a Pop-up Shop as part of the Employment Pathways curriculum. 69
16 The teacher (me) eliciting a fair price from Year 8/9 students while displaying the first batch of bush medicine. 70
17 Sharmic (centre right) and Henry (centre left) consult the catalogue created for the Utopia Desert Art Pop-up Shop. 71

18 A Year 9 student is selling a bush medicine original artwork and completing the cash transaction while the customer (image cropped) stands alongside her. 72
19 A *Typhonium* plant growing at the base of a river red gum tree (*Eucalyptus camaldulensis*) in an area downstream of the permanent water source. The broad leaves on this plant are unique for a desert ecosystem. 73
20 A student points to the direction the transect line will be laid. 81
21 A yarning circle conducted on the akatyerr growing site, showing the relational processes. Two elders, Jedda Purvis (blue beanie) and Judy Purvis (pink beanie), with the assistant teacher, Louisa Long (red shirt), and the teacher, Nicole Boyd (blue shirt), are interacting during the student-led inquiry. 91
22 Auto-coded themes generated from yarning circles at the akatyerr site. 96
23 The right way to conduct two-way dialogue on akatyerr. 99
24 Student interests in akatyerr knowledge concepts pooled by each student participant and classroom assistant teacher, Sharmic. 104
25 Student interests in akatyerr knowledge concepts related to people, country, and land management contextualised to the Australian Curriculum as question prompts for yarning circles. 106
26 Phases of the Goompi pedagogical cycle (reality, abstraction, mathematics, reflection) adapted for the akatyerr harvest context from Matthews (2006). 111
27 Dried akatyerr – a concrete referent packaged by students. The whole dried akatyerr pictured was harvested by students and staff. 113
28 Yarning circle at akatyerr harvest site: Students yarn with adults during the reality phase of the Goompi measurement pedagogical cycle. 116
29 Example creative act: walking the transect line to locate the quadrat point. 116
30 Example student artefact recording two attributes: length (Quadrat point column) and area (% Cover column). 118
31 Student representation of area of the 1 m^2 quadrat covered in akatyerr vegetation (also mass and capacity) in photographed image of quadrat point 49 (visual representation). 119
32 Student representation of area of the 1 m^2 quadrat covered in akatyerr vegetation at quadrat point 49: no vegetation drawn on the graph paper (visual representation) and graph labelled 0% (numeric representation). 119
33 Example student artefact recording two attributes: length (Quadrat point column) and area (% Cover column) relating to transect line number 3. 120
34 Student representation of area of the 1 m^2 quadrat covered in *Typhonium* (or pretty flower) vegetation in photographed image of quadrat point 39. 121
35 Grid reference for estimating percentage cover (adapted from Deakin University, School of Life and Environmental Sciences, n.d.). 122

36 Example student artefact: Student 1's drawing of the area of the 1 m^2 quadrat covered in *Typhonium* vegetation at quadrat point 39. Student 1 worked in a pair with Student 2. 122
37 Example student artefact: Student 2's drawing of the area of the 1 m^2 quadrat covered in *Typhonium* vegetation at quadrat point 39. 124
38 Example student artefact of a creative act: photographing the area of the 1 m^2 quadrat covered in grass vegetation at quadrat point 37. 125
39 Semi-permanent soak (freshwater discharge) at the akatyerr site: a place to reflect the reality. 126
40 The RATIO pedagogical model phases: Reflect, Act, Trial, Interact, Opportunity. 149
41 Example question to demonstrate disciplinary thinking of area of an array (and procedure). 150
42 Example trial of the Goompi pedagogical cycle – reality, abstraction, mathematics, and reflection (explicitly adapted for the akatyerr quadrat survey; Goompi pedagogical cycle adapted from Matthews, 2006). 154
43 Creating opportunities for students to conduct a quadrat survey. Assistant teacher Sharmic is pictured standing, and I am pictured crouching with the quadrat survey instrument held up to demonstrate to students. 156
44 Map illustrating the location of Alyawarr, Eastern Anmatyerr, and neighbouring languages. 172
45 Aligning the quadrat frame at a quadrat point demonstrated by Student 3 (left). Placing the quadrat frame along transect line demonstrated by Student 4 (right). 177
46 Culturally inclusive Year 8 work sample – solving problems relating to the volume of prisms. 181
47 Aerial image of akatyerr harvest site with parallel transect lines (red lines). 182
48 A section of the aerial image of akatyerr harvest site transformed and represented to solve area of rhombus (by applying trigonometry formula as Area = $S^2 \sin A$ or Area = $S^2 \sin B$). 183
49 Geometric reasoning (congruent angles) and solving area of a rhombus (Area = $S^2 \sin B$). That is, if S (side length) is 50 and angle <ABD is 100° then Area = $S^2 \times \sin(100)$. 183

Tables

1 Indigenous culture content in NTILC curriculum for first languages maintenance. 17

2 Curriculum adjustments for contextualising measurement learning to sustainable wild akatyerr harvest: Integrating Indigenous languages and cultures, science cross-curriculum priorities, and mathematics for Years 8–9 students. 75
3 Summary of each trial of the Goompi pedagogical cycle adapted for contextualising measurement attributes to akatyerr harvest. 127
4 Rich description of the participants in each participant group. 137
5 Summary of the influences filtering through my narrative of my lived experience. 142

Abbreviations

ACARA	Australian Curriculum, Assessment and Reporting Authority
AIHW	Australian Institute of Health and Welfare
AITSL	Australian Institute of Teaching and School Leadership
ANFAB	Australian Native Food and Botanicals
ATSIMA	Aboriginal and Torres Strait Islander Mathematics Alliance
BB	Boundless Border (in Yarning Circle Coding)
BLM	Bureau of Land Management
CANZUS	Canada, Australia, New Zealand, and United States
CAYLUS	Central Australia Youth Link Up Services
CLC	Central Land Council
ICIP	Indigenous Cultural and Intellectual Property
IEK	Indigenous Ecological Knowledge
NAPLAN	National Assessment Program – Literacy and Numeracy
NTBOS	Northern Territory Board of Studies
NTDOE	Northern Territory Department of Education
NTG	Northern Territory Government
NTILC	Northern Territory Indigenous Languages and Cultures
OECD	Organisation for Economic Co-operation and Development
PISA	Programme for International Student Assessment
QUT	Queensland University of Technology
RTO	Registered Training Organisation
SCRGSP	Steering Committee for the Review of Government Service Provision
SS	School Site (in yarning circle coding)
STEM	Science, Technology, Engineering, Mathematics
TESOL	Teaching English to Speakers of Other Languages
UNDRIP	United Nations Declaration on the Rights of Indigenous Peoples
VET	Vocational Education and Training

About the Author

Nicole Boyd
is a descendant of a large family from Broken Hill, New South Wales, on her mother's side. Her grandfather hunted kangaroo and shared it with large family groups extending to Wilcannia. Nicole is also a descendant of a small rural family from Port Fairy, Victoria, on her father's side. Her grandmother, a great horse rider, is the youngest daughter of a farrier born in Bendigo. Her grandfather, a travelling showman, inspired Nicole when she first took to the Wistari Room stage at Heron Island (Great Barrier Reef) resort. Nicole, as an early-career scientist, related stories of marine turtle nesting behaviour to resort guests. The marine turtle interpretation sessions at Heron Island and other turtle nesting beaches stimulated her lifelong interest and career in STEM education. She is a sibling to four brothers and one sister, and alloparent to four nieces and five nephews. Nicole is married to Paul Evans, the love of her life, whose depth of commitment on this journey of life is unfathomable.

CHAPTER 1

Introduction

In 2011, I began my first teaching role at an all-Aboriginal girls secondary boarding school, Worawa Aboriginal College. Founded by Aboriginal poet and playwright Hyllus Maris, this independent school in regional Victoria combines Aboriginal culture and mainstream Australian culture in its curriculum (Hamilton, 2019). The school is located on the fertile soils that were once farmed by members of the Kulin Nation at Coranderrk, "the Woiwurrung name for the flowering 'Christmas Bush' (*Prostanthera lasianthos*) native to the area" in the picturesque Yarra Ranges (Nanni & James, 2013, p. 8). The school encourages students to envision and navigate pathways to confident and meaningful lives. Worawa's Principal and Executive Director, Dr Lois Peeler AM, is a Yorta Yorta and Wurundjeri elder. Aunty Lois continues the vision of her sister, Hyllus Maris, who, in her oratory at the school's inception in 1983, explained:

> in this, the first Aboriginal school in Victoria, the educational curriculum has been specially designed to suit Aboriginal students to bring them to their full potential. Aboriginal culture will be imparted not only as a school subject in each class's timetable but as an integral part of everyday life at the school. (Hooley et al., 2021, p. 20)

Aboriginal culture was imparted in mathematics lessons particularly through the "formation of partnerships with a range of different school and community organisations" (Hooley et al., 2021, p. 31) while I was teaching at the school. In 2012, I met Quandamooka mathematician and one of the founding members of the Aboriginal and Torres Strait Islander Mathematics Alliance (ATSIMA) Professor Chris Matthews. Chris was involved in the Maths and Indigenous Learners Closing the Gap: Make It Count project. As a Worawa mathematics teacher, I helped run a maths "Big Day Out" event – a play on the name of a music festival here in Australia. At our maths "festival", Chris demonstrated how to solve simultaneous equations from Year 9 students' investigation of shooting hoops from different positions on the basketball court. From that moment on, I was inspired by Chris's leadership in transforming mathematics – offering innovative pedagogy (see Matthews, 2009) to provide opportunities for all Aboriginal students to participate successfully in mathematics.

Many boarding students living far away from the sandplains of the Central Desert region of the Northern Territory were often homesick for their country

 | DOI:10.1163/9789004682757_001

while at Worawa. I did not know the importance of students' connection to country for their identity. I became curious to know what it was like to live and teach alongside students and their families, so in 2014 I applied for a teaching position in a remote Alyawarr-speaking community and moved there. At night I would hear the church choir singing, and during the day, the same elders who led the church worked tirelessly helping to teach a colonising curriculum. Here, Aboriginal culture was expected to be left at the classroom door. After teaching and learning alongside both Alyawarr and Anmatyerr assistant teachers at small remote Northern Territory schools (with three teachers), I accepted a role as both secondary teacher and mathematics education researcher in a Central Desert community, aptly called Utopia. The school principal ensured new teachers were advised by the local elder and chair of the school council at the beginning of each term. Here, quietly spoken elders would speak to the non-Indigenous teaching staff and instruct us on ways to proceed, respectfully.

Utopia Eastern Anmatyerr and Alyawarr elders recommended:

> *Mwerr kwerltety-rnem aweyel, akaltyerreyel Ampernem inkwerelhem kwerlwarl alheyel.*
> It's good when school teachers can listen and understand our way. The students are happy to come to school when the teachers understand their way. (Moore, 2014, p. 12)

Understanding "their way" requires teachers to create innovative approaches to teaching and learning. Here, I began to envision curriculum that would incorporate both Aboriginal culture and mainstream Australian culture. The principal of the school encouraged all non-Indigenous staff to attend Alyawarr language lessons to assist with this. Also, I was encouraged to coordinate "both-ways" science excursions or *bush trips* (Ober, 2009; Ober & Bat, 2007). On 25 October 2016, partnered with the Central Land Council (CLC), Central Australia Youth Link Up Services (CAYLUS), Wanta Aboriginal Corporation, middle years students, elders, assistant teachers, and formally qualified teachers, we began our first country visit to a culturally significant rock hole. The question I was continually guided by was: how can we respectfully integrate Indigenous knowledge into mathematics curriculum?

> I began the work that I could do towards systemic social change in school curriculum and pedagogy. By examining my own positioning and lived story working and studying in relation to Eastern Anmatyerr and Alyawarr people and country, for instance, I came to know how to decolonise and

> transform my teaching and research practices. (Nicole Boyd, field notes, November 29, 2019)

I looked for examples of where a teacher's improved sense of self-efficacy in meeting the needs of culturally and linguistically diverse students was likely to improve the "achievement gaps" in their students (Pledger, 2018, p. 108). While I did not measure teacher efficacy as part of my research, nor students' achievement gaps, I sought to find ways to understand how to include Indigenous perspectives in curriculum and pedagogic design choices. A more culturally inclusive and socially just pedagogy situates the teacher as being responsible for and having agency "to resist dominant norms" in pedagogical decision-making (Baggett et al., 2020, p. 160). In my study, I understood teacher self-efficacy as the capacity to be flexible so as to appropriately link mathematics curriculum and pedagogic design choices to Indigenous identities, knowledge, and values through forging a strong partnership with Indigenous assistant teachers.

A study on teacher self-efficacy in the USA revealed teachers themselves need to be "actively seeking and engaging in educational experiences that improve their ability to interrupt patterns of inequity" and develop patterns that meet the needs of their students (Pledger, 2018, p. 104). That is, it is necessary to provide opportunities to empower teachers to adjust their teaching behaviour. This is because "simply hearing rationale, suggestions, or motivations about why engaging in a particular practice could be beneficial" (Pledger, 2018, p. 106) does not give teachers the opportunity to assess their own professional growth needs and progression towards culturally responsive education. Similarly, in remote Australian contexts, a teacher's efforts to continually develop their cultural competence will result in increased opportunity for and access to education for their students.

Australian teachers with a "low level of critical awareness of issues of race, culture and power in teaching and learning" may fail to provide appropriate levels of academic challenge for their students (Carter Andrews, 2014, 3:40). Teachers' lack of awareness of their position within an inequitable education system may contribute to perpetuating disparities for racially and socioeconomically diverse students. As a result, my study recognised the urgent need to determine "how the voices of the Indigenous can be made the privileged ones" (Olsen, 2017, p. 206) in education. However, the need to privilege Indigenous voices in education comes because most efforts to address the privileged position of dominant (white) educators focus more on "self-awareness" of the dominant culture rather than "social change" (Miller, 2013, p. 93). Education research efforts have failed to address "how [teachers] build capacity to disrupt existing networks of power in the contexts in which they work" (Miller, 2013, p. 93) to subvert power relations in the classroom. Western teachers are complicit in a silent transmission of the dominant cultural norms when they unquestioningly enact exclusionary teaching practices in remote Indigenous classrooms.

Transformative changes I made towards equitable mathematics learning began with examining the mismatch between the teaching style prescribed for a lesson in a remote classroom (which can be tightly scripted) and the learning needs of my students. Many students in the remote schools I taught in actively resisted a didactic approach to teaching and learning by refusing to attend a classroom – they could not see themselves reflected in the curriculum. When all the details of a lesson are decided without considering the cultural capital that the students bring to school, students are denied an opportunity to develop greater agency in their learning through tailoring lessons to reflect their interests. Indigenous leadership in my local school community continued to reinforce the need to place Aboriginal culture at the centre of education practices.

1 Apmer Akely-Akely

Apmer Akely-akely was the local community board for the Stronger Communities for Children program in the Utopia region, Northern Territory (Australian Indigenous Governance Institute, 2020) at the time of my study. Esau Nelson, chair of the Apmer Akely-akely board when interviewed for the Indigenous Governance Awards in December 2018, described how teaching and learning Alyawarr language is essential so

> future generations will learn to practice our culture – just as we do. That is what we want – to keep our home strong, to keep our place strong. (Reconciliation Australia, 2019, 1:38)

Esau asserted his political authority for the Eastern Anmatyerr and Alyawarr people in Utopia through meaningful nation-building conversations within Apmer Akely-akely. He advocated for self-determination to continue "social relations and cultural practices" in school bush trips (Cornell, 2006, p. 22). Esau, as the chair, defined a sense of a "shared social identity" which encouraged members of the group to share individual opinions that were transformed into "social facts" which "rendered beliefs" adequate as the motivation for "action" (Reicher et al., 2018, p. 129).

The sense of a "shared social identity" based on "a commonality of concerns", such as to continue to practice culture, led members of the group to act to achieve a "sense of a shared reality" (Reicher et al., 2018, p. 129). This resulted in greater participation in school bush trips. For instance, we conducted excursions every term to find and collect amern mwerr (good bush foods). By "walking country" and telling "stories of country", Eastern Anmatyerr and Alyawarr community exercised their right to "express and strengthen" their connection to Utopia lands (Janke, 2021, p. 236). Conducting a cultural practice was a way to ground pedagogy in the aspirations of Indigenous leaders for intergenerational transfer of knowledge. The focus of a bush trip could include finding and collecting ngkwarl (sugarbag), anaty (potato), arwngerrp (bush turkey), aherr (kangaroo), arlewatyerr (goanna) irreyakwerr (onion), and akatyerr (desert raisin), depending on seasonal availability.

Esau's nation (re)building success could be attributed in part to his ability to build connections between his identity and the group identity, which is "informed by common norms, values, beliefs and goals" to continue to practice culture (Reicher et al., 2018, p. 129). What makes some Indigenous nation (re) building efforts succeed and others fade (in Canada, Australia, New Zealand, and United States [CANZUS] nations) is partly due to "the mismatch between Indigenous demands for self-determination and state programs to address socio-economic disadvantage" (Cornell, 2006, p. 1). The source of "inter-group tension" common to each CANZUS nation is "the rights of Indigenous peoples to govern themselves in their own ways and to shape their relations with encompassing societies in ways of their own choosing" (Cornell, 2006, p. 6). The 21st century common concern among CANZUS nations is the "rights to self-determination" (Cornell, 2006, p. 6).

1.1 *Finding Ngkwarl (Sugarbag)*

As I lived and worked in community, I could promote nation (re)building actions – for instance, by accepting an invitation to participate in cultural practices such as finding sugarbag (ngkwarl) on my weekend. My husband and I were invited to share the experience of collecting ngkwarl with families in one of the homeland communities. I became aware of the significance of this cultural practice (see Figure 1) when learning about finding ngkwarl on a bush trip conducted as part of the Indigenous Languages and Cultures curriculum (Northern Territory Department of Education [NTDOE], 2017a).

After chainsawing the bloodwood tree, Clayton Hunter, a member of the school staff and emerging Utopia Ranger Group, finds the sugarbag and tells the families: "Look there! They found lots of sugarbag! We gotta sugarbag here – inside – plenty. Yeah". After sharing the sugarbag with all the members of

FIGURE 1
Ngkwarl (sugarbag) is gently coaxed out of the bloodwood tree with a probing stick
PHOTOGRAPH BY TIM WARE

the community attending the bush trip, he explains that the children "listen to some stories from womens about country".

In this cultural practice and the conversation above, I could see how important it was for families to show children how to find the sugarbag inside the bloodwood tree and tell stories to teach how country is understood. The stories woven into the lesson on collecting sugarbag reinforced a student's sense of "belonging" as students could see themselves in the school curriculum – in the stories and the cultural practices (Giorcelli, 2020). Being involved in cultural practices such as collecting sugarbag, I could see how "the interests" of remote Indigenous students and the "use of real world ... approaches" were more likely to engage learners (Halsey, 2018, p. 31). Further, I began to envision how merging epistemologies could elicit a new way of constructing knowledge that involved an exchange of languages, cultural content, and experiences that could make learning deeply personal and more meaningful.

Australian Government Indigenous education reforms have de-incentivised remote secondary education options. Indigenous students from remote areas have been encouraged to relocate to regional areas as the "watered down curriculum" delivered in remote schools was considered inadequate for secondary education (Wilson, 2014, p. 136). However, most communities consulted in the Northern Territory Review of Education Service Provision in remote and very remote schools (referred to as Homeland Learning Centres) preferred "local secondary education options" (White & Berta, 2019, p. 44).

Families who prefer to have their children learn on their country (in remote contexts) are even further disadvantaged if they send their children to local schools since financial resources have been funnelled away from remote and into regional areas (Wilson, 2014). Predictably, there have been no improvements towards closing the gap in numeracy achievement for secondary Indigenous students living in the Northern Territory during the past decade (Australian Government Department of Prime Minister and Cabinet, 2020, p. 49). My re-imagining of remote mathematics education began with the women harvesters attending forums to discuss how to strengthen their industry.

1.2 *Wild Akatyerr Harvest*

Bush food forums were held locally in Utopia during the cool months in 2017 and 2018. Eastern Anmatyerr and Alyawarr harvesters, middle years students (Years 8–9), their assistant teachers and I attended and participated in these forums. During discussions, harvesters identified intergenerational transfer of knowledge relating to wild akatyerr harvest as a priority for the local bush food industry (CLC, 2017). Consideration of this need for further development of knowledge transfer was reinforced by bush food industry expert and Ngangiwumirr/Arrernte spokesperson Rayleen Brown:

> The women [harvesters] are saying the only way we can teach our young people about bush foods and medicines is through language, but there are no supported bi-lingual curriculum in the schools any more. What they'd like to see, what I'd really like to see, is a whole unit of work based around secondary school kids going out to work with the ladies, learning harvesting techniques, management techniques, getting kids ready to go on to be rangers, looking after country, knowing the seasons for bush foods, when the wattles are ready, when the akatyerr grows. (Brown, as cited in Finnane, 2018, paras. 41–42)

I chose this edible plant, akatyerr, to contextualise mathematics learning, specifically the measurement strand of mathematics, since it is considered one of the most valuable bush food plants in the Central Desert region of the Northern Territory. Dried and fresh wild akatyerr fruit (see Figures 2 and 3) are collected and sold by many Indigenous people living in the Central Desert (Alyawarr speakers from Ampilatwatja et al., 2009, p. 2).

However, Eastern Anmatyerr and Alyawarr harvesters indicated the price for their yield is unfair given the work involved in harvesting, which can take "three day must be, four day, go there, come back afternoon" (Ruby Morton and

FIGURE 2
Wild akatyerr fruit (*Solanum centrale*)
PHOTOGRAPH BY DAVID ALBRECHT

FIGURE 3
Bucketful of dried akatyerr
PHOTOGRAPH BY NICOLE BOYD

Lucky Morton, as cited in Finnane, 2018, para. 12). The harvest of, for instance, "two bucketsful of akatyerr" can be traded for $300 (Finnane, 2018, para. 12).

Decolonising mathematics education involved embedding the concerns of the community to maintain this important plant food at significant harvesting sites for future generations. The transfer of cultural knowledge promoted sustainable wild akatyerr harvest by considering how to address threats such as introduced perennial grasses. In consultation with community members, I obtained consent to learn appropriate traditional cultural content on akatyerr when teaching sustainable harvest protocols and measurement pedagogy.

1.3 *The Utopia Eastern Anmatyerr and Alyawarr Community Context*

The community where the study was conducted was in a Central Desert region of the Northern Territory. This arid zone region covers an area of 3,500 square kilometres of sandy red soil. It is home to approximately 1,000 Utopia Eastern Anmatyerr and Alyawarr people in 18 homeland sites adjacent to the Grandplover (a pseudonym) River (Northern Territory Government [NTG], 2016). The homeland sites are near small semi-permanent freshwater "soaks" (White et al., 2000, p. 23). In arid Australia, soaks are places where "non-saline groundwater discharge" intermittently emerges at the surface and are important "cultural, spiritual and material resources" for local Utopia Eastern Anmatyerr and Alyawarr people (White et al., 2000, p. 23). School staff members drive buses to transport students between homeland sites and the five homeland school communities (four primary and one high school). Walyarra (a pseudonym) is the service centre for all the homeland communities and consists of an aged care centre, Walyarra store, shire council office, police station, undercover basketball court and youth drop-in centre, Australian Rules football oval, the Walyarra High School, and teacher housing (NTG, 2016).

The Eastern Anmatyerr and Alyawarr community have traded akatyerr wild harvest collected from this region of the Central Desert for over 40 years. In 2017, the women harvesters who convened at a seed-collecting workshop were considered the backbone of the industry (CLC, 2017). Discussions at bush food forum events had raised community concerns on the environmental threats to the sustainability of their industry. Threats such as Buffel grass (*Cenchrus ciliaris*), an invasive, "highly competitive grass" species, were highlighted as they limit akatyerr plant distribution (NTG, 2018, p. 10).

Historically, the Utopia people have never been fully dispossessed because this region was never either a mission or a government settlement (NTG, 2016). Traditional owners and caretakers of the land were granted non-exclusive native title rights in 2014 (National Native Title Tribunal, 2014). Indigenous governance over community matters occurs locally at a small council and through a large, centralised land council. The Walyarra school site is administered by a school principal. The principal adopted a strategy to increase local employment at the school, which involved "representatives from the main family groups" (Douglas, 2011, p. 24). The school's management was informed by regular school council meetings where attending school council members had a say in the future administration of the school site.

Utopia Eastern Anmatyerr and Alyawarr school community members knew me through my role as a teacher and as a member of the community. I acted with cultural sensitivity and built respectful relationships among Utopia Eastern Anmatyerr and Alyawarr community members. During 2017–2019, I worked alongside assistant teacher Sharmic, a Utopia woman who was knowledgeable in traditional acts such as painting and harvesting bush medicine (e.g., *Eremophila dalyana*). *Eremophila dalyana* is a significant plant species found in parts of arid Australia where the understory of predominantly Georgina gidgee

(*Acacia georginae*) vegetation is "dominated by grasses" (White et al., 2000, p. 36). Throughout the 3-year teaching and learning period of the study, Sharmic and I developed an authentic relationship by exchanging an understanding of ways to contextualise, for instance, locally created artworks and bush medicine to measurement curriculum and pedagogic practice.

2 Re-imagining Curriculum and Pedagogy

Re-imagining curriculum and pedagogic practice is of critical importance as education policy and practice have been unable to respond adequately to "entrenched intergenerational socio-economic disadvantage" associated with geographic remoteness, isolation, and an "unresolved historical legacy of institutionalized racism" (Fogarty et al., 2015, pp. 5–6). "Socio-cultural determinants" such as these constrain equitable access to education for very remote and remote Indigenous students compared to other Australian students (Fogarty et al., 2015, p. 7). Figure 4 illustrates known ways remote Australian teachers can deliberately replace unconscious bias revealed by deficit thinking and associated limiting perceptions of potentiality of remote Indigenous students. Replacing deficit thinking with a narrative of strength promotes successful outcomes for Australian Indigenous students (Sarra, 2011). The praxis illustration (Figure 4) depicts how discourse, and perceptions of potentiality

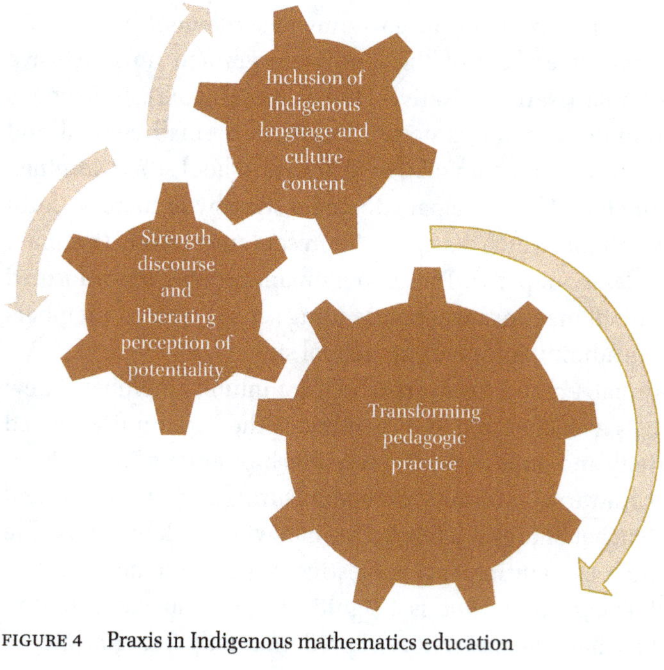

FIGURE 4 Praxis in Indigenous mathematics education

linked with Aboriginal identity, can be disrupted. The praxis denotes the reflections and actions directed at pedagogic decision-making to transform pedagogic design.

Figure 4 depicts how praxis (Freire, 1972) was understood in my study. Re-imagining what school knowledge and practices look like promotes decision-making that reflects and directs action to transform pedagogic practice. For instance, my study was re-imagined so measurement curriculum and pedagogic design included Indigenous language and culture content, a strength discourse, and a positive perception of the "potentiality of educational outcomes and performance of Indigenous students" (Fogarty et al., 2015, p. 8). Two processes depicted in Figure 4 were essential for re-imagining and transforming pedagogic practices. First, promoting a liberating perception of student potential involved examining the prevailing paradigm of school mathematics in my classroom. A liberating perception of potentiality was sought by examining the interplay between the students, educators, and mathematics curriculum knowledge in school-based teaching and learning experiences (Scherer & Steinbring, 2006). That is, I acted to examine the "interaction processes" between the students, their teacher and assistant teacher, and measurement curriculum knowledge (Scherer & Steinbring, 2006, p. 160) in Employment Pathways numeracy learning (NTDOE, 2021). Examining interactions and, notably, replacing pervading deficit discourses with a strengths discourse occurred through setting reciprocal expectations. That is, the students were expected to "actively produce" their own measurement knowledge, and the students expected that their teachers would encourage them "to communicate their own ideas, interpretations and understanding of the mathematical knowledge" learned (Scherer & Steinbring, 2006, p. 174).

The second process depicted in Figure 4 involved examining the possibility of subjugation of a student's identity with English language and culture instruction as the prevailing classroom practice for teaching mathematics knowledge. Transformation of pedagogic practice can involve the inclusion of students' home language and culture in the language used for mathematics instruction. Deficit discourses permeate examinations of Indigenous education (Australian Government Department of Prime Minister and Cabinet, 2020; Wilson, 2014). Deficit discourses, however, do not acknowledge the "historical legacy of institutionalized racism" that remains in Indigenous education structures and practices (see Fogarty, 2012; Fogarty et al., 2015, p. 7).

The assumption that Indigenous students will achieve statistical parity with non-Indigenous students by a "replication of 'normal' education practices" is flawed (Fogarty et al., 2015, p. 5). One flaw in mainstream education is that normalised practices fail to "recognize and respond to the complex socio-cultural determinants" associated with Indigenous education levels of achievement

(Fogarty et al., 2015, p. 7). However, to disrupt the political imperative to examine statistical disparity, teachers can focus on creating positive Indigenous identity and building Indigenous students' sense of self-efficacy (Fogarty et al., 2015; Matthews, 2012; Sarra, 2011). For this reason, measurement pedagogic design that connects with the lived reality of Indigenous students deliberately reframes their identity within a strength discourse and enfranchises cultural production such as food, language, and cosmology or knowledge of "deep spiritual relationship[s]" (Moggridge & Mihinui, 2020, "Traditional lore and customs" section, para. 1).

Curriculum that "engages with and values" Indigenous knowledge supports a positive Indigenous identity (Lingard, 2005, p. 175). Further, curriculum that is relevant is more likely to involve a "substantive sense of satisfaction with, and a psychological investment in" the learning students are participating in (Munns & Woodward, 2006, p. 194). This is important since many communities that speak the words belonging to Utopia Eastern Anmatyerr and Alyawarr language groups have experienced significant barriers to language maintenance (Marmion et al., 2014) since funding ceased with the "abolition of bilingual education by the Northern Territory Government" (Karidakis & Kelly, 2018, p. 121; see also Aboriginal and Torres Strait Islander Social Justice Commissioner, 2014). The Northern Territory Board of Studies (NTBOS), however, recognises the need to maintain "cultural knowledge and language as a basis for learning 'both-ways', or both Indigenous and non-Indigenous systems of knowledge as additive multilingualism and multiculturalism" (NTBOS, 2017, p. 32; see also Ober, 2009; Ober & Bat, 2007).

Cultural inclusivity in education refers to establishing learning environments that are inclusive for culturally diverse students (Dreamson et al., 2017; Gargiulo & Metcalf, 2012). National statements on inclusivity in the Australian Curriculum are broadly based on the assumptions "that all students can learn and that every child matters" (National Curriculum Board, 2009, p. 8). In reality, however, a pervading belief that the "learning capacity of Indigenous children is somehow inferior" (Davis & Grose, 2008, p. 3) underpins diminishing expectations that teachers and Indigenous students have for attaining parity with non-Indigenous students (Sarra, 2005, 2011).

Cultural inclusivity, for my study, refers to creating new learning opportunities for Indigenous and Western knowledge concepts to be represented in measurement curriculum and pedagogic practice. That is, the study promoted students' "equal participation" in measurement learning by "reflecting their cultural background" in the Akatyerr investigation (Dreamson et al., 2017, p. 948). The inclusion of local cultural content aligns with the imperative for mathematics curriculum to be "inclusive of the experiences and knowledge of all students, including marginalised ones" (National Curriculum Board, 2009, p. 6; see also Education Council, 2019; Guenther & Osborne, 2020). The inclusion of local language and cultural content in measurement curriculum and pedagogic practices promotes a more equitable assessment of learning. Moreover, the inclusive imperative aligns with the Northern Territory Government's framework for inclusion (NTDOE, 2019) and the framework for teaching "Indigenous languages and cultures" (NTBOS, 2017, p. 6). Equitable access to measurement learning involves the "crafting of individual curricula, honouring context and culture, nurturing relationships with families, and continually interrogating practice to ensure all are deriving benefit" (Ryan, 2016, para. 7).

Understanding that differences exist in the cultural and social norms of the students and those of the mathematics classroom practices in remote school contexts is critical in creating inclusive learning opportunities. Although creating opportunities for "dialogue" about students' lived experiences cannot dismantle "systemic issues of power and privilege" in education (Buckley & Quaye, 2016, p. 1121), nevertheless, hearing student voice supports a strengths discourse in the school context. The decisions that lead to the inclusion of students' experiences and interests in the curriculum can be guided by the aspirations of the students' families and broader community. Curriculum and pedagogical decision-making can be grounded in social justice (Lingard, 2005) to prevent social and cultural differences becoming the reason to abandon some students to "low expectations" that limit their development (National Curriculum Board, 2008, p. 9).

Despite national curriculum initiatives to embed Aboriginal cultural content across the curriculum (Australian Curriculum, Assessment and Reporting Authority [ACARA], 2020a; Davis & Grose, 2008; Riley & Genner, 2011), there is limited understanding of the role that cultural inclusivity plays in students' engagement with mathematics curriculum (Jorgensen, Grootenboer, & Niesche, 2013). The role that cultural inclusivity of engendered social norms, such as arranging students with consideration to kinship, may play in student engagement with and access to mathematics is not widely understood. The practice of group work or reporting back, for instance, can be problematic for remote Indigenous students, as their cultural protocols relating to kinship or "skin groups" (Jorgensen et al., 2010) preclude some students from working and speaking together. This issue was of critical importance in my study.

For these reasons, my study considered how community voices could be heard (Guenther et al., 2017). More specifically, this involved considering what steps a non-Indigenous teacher-researcher could take to hear and amplify the voices so the research could benefit all community members (Guenther et al., 2017).

2.1 *Sharing Stories on Country*

Pedagogic design that incorporates "practice in real or role play format" – for instance, through sharing stories (see Figure 5) – improves "the likelihood of [students'] self-efficacy development" (Pledger, 2018, p. 107). "Personal judgments of one's capabilities" developed in conversation between teachers, students, and elders can lead to successful "educational performance" (Zimmerman, 1995, p. 202; see also Bandura, 1977). A student's personal judgement of capability develops as they see themselves, their interests, and their experiences in curriculum and pedagogic practices.

Student understanding is deepened through the practice of sharing stories about the interrelationships of people, country, and creation time in order to understand a local context (Dobson et al., 2008; Merne Altyerre-ipenhe Reference Group et al., 2011; Turner, 2005) and through making connections with the school-based language and practices. When asked by students how to care for this country, one elder answered that "caring for country is knowing which people are about to do what in relation to sacred sites" (Yarning circle BB1,[1] June 13, 2019).

Re-imagining what school knowledge and pedagogical practices look like to include the traditional lore or "rules and proper ways to behave on and treat Country" associated with managing country is of critical importance (NTDOE, 2017b, p. 17) since

> Indigenous people are bound by their traditional lore and customs. The stories, songs and history relate to where the people have come from

FIGURE 5 Utopia Alyawarr ranger (fifth from left, wearing a yellow shirt) sharing stories about caring for his country
PHOTOGRAPH BY TIM WARE

> (creators, cultural heroes and ancestors) and where they are going. This sets a deep spiritual relationship with the land, sky and waters. … Generations of Indigenous people are taught by their elders to respect the laws of the land. They believe that if the lores are broken, traditional and modern justice may be imposed. (Moggridge & Mihinui, 2020, "Traditional lore and customs" section, paras. 1, 3)

I sought to understand how to create space in the curriculum and pedagogic practice for intergenerational transfer of knowledge relating to akatyerr to occur, while attending to and maintaining traditional lore and customs.

2.2 *Intercultural Spaces – Respect for Traditional Lore*

Yolŋu teachers describe the intercultural space by their metaphor, Ga<u>n</u>ma. Ga<u>n</u>ma is an imagined (and real) place where saltwater people (Yolŋu) and freshwater people (Balanda) come together and mix and mingle (Lloyd et al., 2016, p. 1). The metaphor of mingling explains how ideas are treated in the intercultural space. New ideas are constructed through Ga<u>n</u>ma, at the site where Western and Yolŋu mathematics knowledges merge (Lloyd et al., 2016). Mathematics, like all knowledge for Yolŋu people, cannot be understood without considering country and kinship (Lloyd et al., 2016).

Similarly, for Utopia Eastern Anmatyerr and Alyawarr people, knowledge of akatyerr is also understood in relation to the interconnection of people, country, and creation time (Dobson et al., 2008; Merne Altyerre-ipenhe Reference Group et al., 2011; Turner, 2005). However, care needs to be taken so that "too much responsibility" is not expected of young individuals in a community (Moore, 2014, p. 4). This is because "visitors to a community [may] expect the young, educated, English-speaking people in the community to make decisions for the community without consultation with elders" (Moore, 2014, p. 4). A considered approach is needed to create an equitable mathematics education which blends "traditional culture, ceremonial life and western knowledge" (Halsey, 2018, p. 32).

Students guided in conversations with elders can gain knowledge and understanding from traditional lore and leverage from this a new understanding of measurement knowledge. Incorporating differences in ways of knowing when making pedagogic choices, such as the use of the quadrat survey method in my study, provides learning experiences that support students to move between the familiar and the unfamiliar (Aikenhead, 2017; Ober, 2009; Ober & Bat, 2007). The quadrat survey method allowed students to assess the relative abundance of plants found at the site. Quantifying the changes in relative abundance of plants over time is a land management practice used to monitor plant protection efforts (such as exclusion fencing; Philippoff & Cox, n.d.).

FIGURE 6
Constructing new measurement knowledge in an intercultural space (adapted from Northern Territory Department of Education, 2017b, pp. 15–21)

Consultation with and participation of Indigenous elders is required so locally and communally owned Indigenous knowledge relating to place is treated with respect when contextualised to the Australian Curriculum. Pedagogical practices that merge ways of knowing, and which are linked appropriately to community needs and aspirations, can improve a student's sense of potentiality connected to their Indigenous identity at the site where knowledge is created.

Figure 6 depicts how my study re-imagined measurement pedagogic design. The design related the content from the measurement sub-strand of the Australian Curriculum: Mathematics (ACARA, 2020c) to ways of talking from the Northern Territory Indigenous Languages and Cultures (NTILC) framework (NTDOE, 2017b; see also Table 1).

Listening to the voices of the participants was critical to create an intercultural space where both worldviews could be represented (Frawley & Fasoli, 2012). Yarning circles were understood in the study as "meeting[s] of peers or participants of similar status for discussion and exchange of views" (Woolley et al., 2013, p. 92). Yarning circles helped to remove barriers to learning since I could "embrace the role as facilitator of learning" and in doing so create "spaces and time for dialogue" (Davis & Grose, 2008, p. 14) with students, assistant teachers, and community members. In yarning circles, the intercultural space incorporated Indigenous identities and cultural knowledge and values in relation to the community's aspirations for intergenerational transfer of knowledge relating to sustainable akatyerr harvest.

TABLE 1 Indigenous culture content in NTILC curriculum for first languages maintenance

Culture content	Country/land	People/kinship	Natural environment
Band 2	Understand the rules and proper ways to behave on and treat Country Listen to oral histories and stories (in response to students' questions)	Understand kinship and Country Learn ways to look after themselves and each other	Listen to traditional stories about the seasons and how to live on Country; find water and care for Country (in response to students' questions)

SOURCE: TABLE EXCERPTS ADAPTED TO MY STUDY FROM NORTHERN TERRITORY DEPARTMENT OF EDUCATION, 2017B, P. 17

> So, how do you care for country – white fella way? By acknowledging the right people who can talk for country. Have the right questions. Ask how to care for country in relation to who can talk for country, and where the relevant sites are. (Yarning circle BB1, June 13, 2019)

Table 1 indicates how culture content, such as ways of speaking and acting, can be re-imagined and connected to the Australian Curriculum. For instance, Indigenous culture content can be connected to general capabilities such as numeracy, cross-curriculum priorities such as sustainability, and science disciplinary knowledge (ACARA, 2020a, 2020b; NTDOE, 2017b). The connections across the Australian Curriculum aim to "deepen [student] knowledge ... in their home language and culture" and enable students to "feel strong about themselves as speakers and members of the language group" while building connection to school learning (NTDOE, 2017b, p. 3). This enables students to deepen their intercultural understanding of ways of seeing and talking about the world. Drawing on the knowledge of "deep spiritual relationship[s]" was critical, particularly at sacred sites such as at the akatyerr harvest site (Moggridge & Mihinui, 2020, "Traditional lore and customs" section, para. 1). The proper way to behave on country was guided by assistant teacher Louisa in the following excerpt:

> Not going through sacred sites – inadvertently. How people, everyone can learn. Not blundering in – maybe going this way and that – inadvertently. (Yarning circle BB1, June 13, 2019)

Australian teachers' mainstream classroom practice reflects an "incremental and deliberate shift away from localized forms of educational development that acknowledge and include Indigenous wants and needs" (Lingard et al., 2013). The shift has occurred because of a global movement to privatise education. The shift comes from multinational corporations who oversee "the direction of school reform" (Fogarty et al., 2015, p. 4). Similarly, in Australia, education policy reforms focus on statistical measures of assessing teaching and learning effectiveness. The preoccupation with the political imperative of achieving statistical parity has resulted in Australian policy goals that are not connected to the real-life experiences of Indigenous students (Altman & Fogarty, 2010).

Moreover, Australian teacher practice reflects global education reforms that contribute to the de-pedagogising and de-professionalising of teachers (see Ravitch, 2010). Professionalism in the Northern Territory involves creating learning experiences that "affirm and challenge our learners", while teachers are expected to "develop by example the principles of social justice and equity" (Teacher Registration Board of the Northern Territory, 2019). By contrast, within the remote Northern Territory there is a growing perception that pedagogic approaches such as Direct Instruction, which standardise the delivery of literacy and numeracy, are effective because "all details of instruction" are "controlled to minimize the chance of students' misinterpreting the information" (National Institute for Direct Instruction, 2020). Thus, the remote Northern Territory teachers in schools that have mandates to deliver de-pedagogised and de-professionalised or scripted lessons are faced with limited options to improve the quality of mathematics teaching and learning (Ewing, 2011; see also Rigney et al., 2020).

2.3 *Being Responsive: Grounding Pedagogic Design in Students' Lived Realities*

Grounding pedagogical decision-making in a real-world context that is also relevant to students' interests promotes Indigenous middle years students' engagement in measurement learning. Focusing in on the local context can allow students to apply measurement disciplinary knowledge at a site while furthering the community-voiced aspirations in maintaining a site. Additionally, because a local context is place based or "connected with the community in which the school is located" (Halsey, 2018, p. 33), it is more likely to be relevant and of interest to remote students. While the "development of financial and entrepreneurial skills" is considered appropriate for engaging middle years students (Education Council, 2019, p. 8), problems may arise when the context for developing these skills is done without respect for the right Indigenous people have to "maintain, control, protect and develop", for instance, seeds or fruit (United Nations, 2007, Article 31, para. 1).

Curriculum enacted at culturally significant places should be done with sensitivity and with respect for cultural values relating to that place. The *United Nations Declaration on the Rights of Indigenous Peoples* (UNDRIP; United Nations, 2007) recognises that "Indigenous peoples have the right to maintain, control, protect and develop their cultural heritage, traditional knowledge and traditional cultural expressions" including, for instance, "knowledge of the properties of flora and fauna" and "oral traditions" (Article 31, para. 1). In my study, consultation between Indigenous and non-Indigenous staff was essential to co-design lessons that provided high-quality measurement learning while respecting the oral traditions and cultural obligations of people, places, and the plants and animals that relate through kinship to that place (Green et al., 2019).

The mathematisation of a traditional act such as the sustainable wild akatyerr harvest, if planned and implemented by a non-Indigenous teacher without reference to the appropriate cultural authority, could be problematic. For instance, integration of worldviews in "familiar and highly focused specialist contexts" (Frawley & Fasoli, 2012, p. 316) without the appropriate cultural authority may not be able to respond to "new and changing circumstances" (Frawley & Fasoli, 2012, p. 316) that could emerge while visiting an important cultural site for akatyerr harvest. Supporting students to develop new understandings of lived experiences and mathematical ideas when combining knowledge represented from both cultures is one way to provide a culturally safe space where students can strengthen their sense of self while learning mathematics (Lloyd et al., 2016).

2.4 *Cultural Competence*

Developing cultural competence is recommended particularly for non-Indigenous teachers working at Australian Indigenous remote education sites (Moore, 2014). The professional development of remote educators should include learning how to address the potential for cultural conflict (Bishop, 1993) in a mathematics classroom – that is, when differences exist in the cultural and social norms of the students' home or community and school (Dickenson-Jones, 2008). Students may experience discomfort, shame, and loneliness and may disengage from the curriculum if their home and community cultural context are not seen to be valued within the classroom (Dickenson-Jones, 2008). As Luneta (2021) reminds us, for Australian Indigenous students, a classroom "culture of shame is about personal reproach or reproach from other individuals" (p. 339). Shame, or "nyerralth" (Green et al., 2019, p. 397), in a Utopia Eastern Anmatyerr and Alyawarr context is a "social and cultural construct that must be taken care of in order to encourage learners to break free from its bondages" (Luneta, 2021, p. 339).To address the potential for cultural conflict when teaching the

Australian Curriculum, teachers should teach in ways that do not replace students' own values with those representing a dominant, Western worldview (Perso & Hayward, 2015). That is, students' own worldviews should be valued while they access the concepts articulated in the Australian Curriculum.

In Australia, there is often a mismatch between the culture of the teacher (metropolitan) and the culture of remote Indigenous students. The Alice Springs (Mparntwe) Education Declaration suggests that learners should be able to see "themselves, their identities and cultures reflected in the curriculum" (Education Council, 2019, p. 15). The adjustments to improve a teacher's cultural competence in curriculum design and classroom practice require ongoing critical reflection and consultation. A suitable blend of Indigenous and Western knowledge concepts in the curriculum should be responsive to community needs. That is, cultural inclusion in curriculum should be negotiated with sensitivity and a capacity to "understand, interact and communicate effectively" (Perso & Hayward, 2015, p. 1) with Indigenous community members. Additionally, the curriculum content should be "inclusive of the experiences and knowledge of all students" (National Curriculum Board, 2008) and give students a "voice in and influence over" their learning (Education Council, 2019, p. 13). Teachers seeking to find ways to hear the voices of the students, then, need to seek permissions on the cultural appropriateness of contextualising students' interest in their learning.

Although grounding pedagogical decision-making in locally relevant and distinct experiences, environment, and language groups develops a teacher's cultural competence (Michie & Linkson, 1999), sensitivity is required. This is to avoid unintentional harm that may be caused when asking for Indigenous people to answer ill-conceived questions. It is important to "always be listening to Indigenous staff in order to find out how much responsibility they can assume" when making decisions on behalf of their community's distinct interests (Moore, 2014, p. 5). Taking care of cultural obligations may preclude Indigenous staff from answering questions as they may not have the cultural authority to speak on behalf of their community. Therefore, it is of critical importance to seek information on who is an appropriate person to ask and validate the cultural authority of those with the responsibility to answer.

Note

1 BB1 is a code for the first yarning circle group at the Akatyerr harvest site; BB stands for "boundless border".

References

Aboriginal and Torres Strait Islander Social Justice Commissioner. (2014). *Social justice and Native Title report 2014*. Australian Human Rights Commission. Retrieved January 11, 2023, from https://www.humanrights.gov.au/sites/default/files/document/publication/SJNTR%20FINAL.pdf

Aikenhead, G. (2017). A 21st century economic, educational and ethical mathematics curriculum policy. *The Mathematics Enthusiast, 14*(1–3), 563–574. Retrieved January 11, 2023, from https://scholarworks.umt.edu/tme/vol14/iss1/29

Altman, J., & Fogarty, W. (2010). Indigenous Australians as "no gaps" subjects: Education and development in remote Indigenous Australia. In I. Snyder & J. P. Nieuwenhuysen (Eds.), *Closing the gap in education? Improving outcomes in southern world societies* (pp. 109–128). Monash University Press.

Alyawarr speakers from Ampilatwatja, Walsh, F., & Douglas, J. (2009). *Angka Akatyerr-akert: A desert raisin report*. Desert Knowledge Cooperative Research Centre. Retrieved January 11, 2023, from https://www.nintione.com.au/resource/DKCRC_Angka-Akatyerr-akert_A-Desert-raisin-report.pdf

Australian Curriculum, Assessment and Reporting Authority. (2020a). *Australian curriculum cross-curriculum priorities: Aboriginal and Torres Strait Islander histories and cultures*. Retrieved January 11, 2023, from https://australiancurriculum.edu.au/f-10-curriculum/cross-curriculum-priorities/aboriginal-and-torres-strait-islander-histories-and-cultures/

Australian Curriculum, Assessment and Reporting Authority. (2020b). *Australian curriculum: Learning continuum of numeracy: Using fractions, decimals, percentages, ratios and rates (Level 5)*. Retrieved January 11, 2023, from https://bit.ly/3GzjB4i

Australian Curriculum, Assessment and Reporting Authority. (2020c). *Australian curriculum: Mathematics (Version 8.4) – Year 8 level description*. Retrieved January 11, 2023, from https://bit.ly/3GB8zM7

Australian Government Department of Prime Minister and Cabinet. (2020). *Closing the gap report 2020*. Commonwealth of Australia. Retrieved January 11, 2023, from https://ctgreport.niaa.gov.au/sites/default/files/pdf/closing-the-gap-report-2020.pdf

Australian Indigenous Governance Institute. (2020). *Exceptional governance: Stories of success from the 2018 Indigenous Governance Awards*. Retrieved October 12, 2022, from https://aigi.com.au/resource/exceptional-governance-stories-of-success-from-the-2018-indigenous-governance-awards/

Baggett, H. C., Dunn, A. H., & Sondel, B. L. (2020). Critical teacher responsibility in tumultuous times: Engaging in a community of practice. In L. Parson & C. Ozaki (Eds.), *Teaching and learning for social justice and equity in higher education* (pp. 157–178). Palgrave Macmillan. https://doi.org/10.1007/978-3-030-44939-1

Bandura, A. (1977). Self-efficacy: Toward a unifying theory of behavior change. *Psychological Review*, *84*, 191–215. https://doi.org/10.1037/0033-295X.84.2.191

Bishop, A. (1993). *Conceptualising cultural and social contexts in mathematics education* [Keynote address]. Contexts in Mathematics Education: Mathematics Education Research Group of Australasia Conference, Brisbane, Queensland, Australia. Retrieved January 11, 2023, from https://merga.net.au/Public/Publications/Annual_Conference_Proceedings/1993_MERGA_CP.aspx

Buckley, J. B., & Quaye, S. J. (2016). A vision of social justice in intergroup dialogue. *Race Ethnicity and Education*, *19*(5), 1117–1139. https://doi.org/10.1080/13613324.2014.969221

Carter Andrews, D. J. (2014, March 11). *The consciousness gap in education – an equity imperative* [Video]. TEDx Talks. Retrieved January 11, 2023, from https://www.youtube.com/watch?v=iOrgf3wTUbo

Central Land Council. (2017, August). Seed collectors want a hand to meet growing bush foods demand. *Land Rights News Central Australia*, *7*(2), 16. Retrieved December 9, 2022, from https://www.clc.org.au/land-rights-news/

Cornell, S. (2006). *Indigenous peoples, poverty and self-determination in Australia, New Zealand, Canada and the United States* (Joint Occasional Papers on Native Affairs, JOPNA No. 2006-02). Native Nations Institute for Leadership, Management, and Policy and Harvard Project on American Indian Economic Development. Retrieved January 10, 2023, from https://www.researchgate.net/publication/238707334_Indigenous_Peoples_Poverty_and_Self-Determination_in_Australia_New_Zealand_Canada_and_the_United_States

Davis, J., & Grose, S. (2008). *Which way? What happens when embedding Aboriginal and Torres Strait Islander perspectives in schools meets the professional standards for teachers and an accountability matrix?* Stronger Smarter Institute. Retrieved January 11, 2023, from https://strongersmarter.com.au/wp-content/uploads/2020/08/Which_Way__by_John_Davis_and_Sharon_Grose_for_Wipce_2008.pdf

Dickenson-Jones, A. (2008). Transforming ethnomathematical ideas in Western mathematics curriculum texts. *Mathematics Education Research Journal*, *20*(3), 32–53. https://doi.org/10.1007/BF03217529

Dobson, P. V., Walsh, F., & Sati, W. (2008). *Anpernirrentye: Relationships between bush foods, creation laws, people, country and all things, illustrated by three plants* [DVD]. Desert Knowledge Cooperative Research Centre.

Douglas, J. (2011). *Learning from country: The value of country visits in remote schools for community engagement and natural and cultural resource management* (Desert Knowledge Cooperative Research Centre Report No. 69). Ninti One Limited. Retrieved January 11, 2023, from https://www.nintione.com.au/resource/NintiOneResearchReport_69_Thevalueofcountryvisitsinremoteschools.pdf

Dreamson, N., Thomas, T., Hong A. L., & Kim, S. (2017). Policies on and practices of cultural inclusivity in learning management systems: Perspectives of Indigenous holistic pedagogies. *Higher Education Research & Development, 36*(5), 947–961. https://doi.org/10.1080/07294360.2016.1263830

Education Council. (2019). *Alice Springs (Mparntwe) education declaration.* Retrieved January 11, 2023, from https://nla.gov.au/nla.obj-2391375250/view

Ewing, B. (2011). Direct instruction in mathematics: Issues for schools with high Indigenous enrolments: A literature review. *Australian Journal of Teacher Education, 36*(5), 65–92. https://doi.org/10.14221/ajte.2011v36n5.5

Finnane, K. (2018, March 2). Bush foods: How can wild harvesters get a piece of the pie? *Alice Springs News, 25*(2). Retrieved January 11, 2023, from https://alicespringsnews.com.au/2018/03/02/bush-foods-how-can-wild-harvesters-get-a-piece-of-the-pie/

Fogarty, W. (2012). Country as classroom. In J. C. Altman & S. Kerins (Eds.), *People on country: Vital landscapes, Indigenous futures* (pp. 82–93). Federation Press.

Fogarty, W., Lovell, M., & Dodson, M. (2015). A view beyond review: Challenging assumptions in Indigenous education development. *UNESCO Observatory Multi-Disciplinary Journal in the Arts, 4*(2), 1–23. Retrieved October 14, 2022, from https://hdl.handle.net/1885/13420

Frawley, J., & Fasoli, L. (2012). Working together: Intercultural leadership capabilities for both-ways education. *School Leadership & Management, 32*(4), 309–320. https://doi.org/10.1080/13632434.2012.708328

Freire, P. (1972). *Pedagogy of the oppressed.* Penguin.

Gargiulo, R. M., & Metcalf, D. (2012). *Teaching in today's inclusive classrooms: A universal design for learning approach* (2nd ed.). Wadsworth Cengage Learning.

Giorcelli, L. (2020, November 23). *Keynote presentation: Day 2 – Dr Loretta Giorcelli* [Video]. Northern Territory Department of Education Leaders' summit 2020: Beyond the horizon. Retrieved January 11, 2023, from https://www.youtube.com/watch?v=uEHMNXbOHZ0

Green, J., Blackman, D., & Moore, D. (2019). *Alyawarr to English Dictionary* (2nd ed.). IAD Press.

Guenther, J., & Osborne, S. (2020). Did DI do it? The impact of a program designed to improve literacy for Aboriginal and Torres Strait Islander students in remote schools. *The Australian Journal of Indigenous Education, 49*(2), 163–170. https://doi.org/10.1017/jie.2019.28

Guenther, J., Osborne, S., Arnott, A., & McRae-Williams, E. (2017). Hearing the voice of remote Aboriginal and Torres Strait Islander training stakeholders using research methodologies and theoretical frames of reference. *Race Ethnicity and Education, 20*(2), 197–208. https://doi.org/10.1080/13613324.2015.1110294

Halsey, J. (2018). *Independent review into regional, rural and remote education: Final report*. Commonwealth of Australia. Retrieved January 11, 2023, from https://www.dese.gov.au/quality-schools-package/resources/independent-review-regional-rural-and-remote-education-final-report

Hamilton, M. (2019). *Walking together to make a difference: A case study of Worawa Aboriginal College.*

Hooley, N., Razoumova, O., & Peeler, L. (2021). Chapter 2 case study: Worawa Aboriginal college. In *Indigenous schooling in the modern world* (pp. 19–38). Brill. https://doi.org/10.1163/9789004505421_002

Janke, T. (2021). *True tracks: Respecting Indigenous knowledge and culture.* NewSouth Publishing.

Jorgensen, R., Grootenboer, P., & Niesche, R. (2013). Teachers' beliefs and practices in teaching mathematics in remote Aboriginal schools. In R. Jorgensen, P. Sullivan, & P. Grootenboer (Eds.), *Pedagogies to enhance learning for Indigenous students: Evidence-based practice* (pp. 75–88). Springer.

Jorgensen, R., Grootenboer, P., Niesche, R., & Lerman, S. (2010). Challenges for teacher education: The mismatch between beliefs and practice in remote Indigenous contexts. *Asia Pacific Journal of Teacher Education, 32*(2), 161–175. https://doi.org/10.1080/13598661003677580

Karidakis, M., & Kelly, B. (2018). Trends in Indigenous language usage. *Australian Journal of Linguistics, 38*(1), 105–126. https://doi.org/10.1080/07268602.2018.1393861

Lingard, B. (2005). Socially just pedagogies in changing times. *International Studies in Sociology of Education, 15*(2), 165–186. https://doi.org/10.1080/09620210500200138

Lingard, B., Martino, W., & Rezai-Rashti, G. (2013). Testing regimes, accountabilities and education policy: Commensurate global and national developments. *Journal of Education Policy, 28*(5), 539–556. https://doi.org/10.1080/02680939.2013.820042

Lloyd, K., Suchet-Pearson, S., Wright, S., Burarrwanga, L., Ganambarr, R., Gunambarr-Stubbs, M., Ganambarr, B., & Maymuru, D. (2016). Morrku Mangawu—Knowledge on the Land: Mobilising Yolŋu mathematics from Bawaka, North East Arnhem Land, to reveal the situatedness of all knowledges. *Humanities, 5*(3), 1–13. https://doi.org/10.3390/h5030061

Luneta, K. (2021). Living culturally responsive mathematics education with/in Indigenous communities. *Research in Mathematics Education, 23*(3), 338–342. https://doi.org/10.1080/14794802.2021.1899970

Marmion, D., Obata, K., & Troy, J. (2014). *Community, identity, wellbeing: The report of the second national Indigenous languages survey*. Australian Institute of Aboriginal and Torres Strait Islander Studies. Retrieved January 11, 2023, from https://aiatsis.gov.au/publication/35167

Matthews, C. (2009). Stories and symbols: Maths as storytelling. *Professional Voice, 6*(3), 45–50. Retrieved December 9, 2022, from https://www.aeuvic.asn.au/sites/default/files/PV_6-3.pdf

Matthews, C. (2012). Maths as storytelling: Maths is beautiful. In K. Price (Ed.), *Aboriginal and Torres Strait Islander education: An introduction for the teaching profession* (pp. 94–112). Cambridge University Press.

Merne Altyerre-ipenhe [Food from the Creation time] Reference Group, Douglas, J., & Walsh, F. (2011). *Aboriginal people, bush foods knowledge and products from central Australia: Ethical guidelines for commercial bush food research, industry and enterprises* (Desert Knowledge Cooperative Research Centre Report No. 71). Ninti One Limited. Retrieved January 11, 2023, from https://www.nintione.com.au/resource/NintiOneResearchReport_71_BushFoodGuidelines.pdf

Michie, M., & Linkson, M. (1999, July). *Interfacing Western science and Indigenous knowledge: A Northern Territory perspective* [Paper presentation]. 30th Australasian Science Education Research Association Conference, Rotorua, Aotearoa New Zealand. Retrieved January 11, 2023, from http://members.ozemail.com.au/~mmichie/interfacing.html

Miller, M. (2013). *Action for change? Embedding Aboriginal and Torres Strait Islander perspectives in early childhood education curricula* [PhD thesis, Queensland University of Technology]. QUT ePrints. Retrieved January 11, 2023, from https://eprints.qut.edu.au/60905/

Moggridge, B., & Mihinui, R. (2020). *Australian & New Zealand guidelines for fresh & marine water quality: Indigenous principles for water quality*. Water Quality Australia. Retrieved January 11, 2023, from https://www.waterquality.gov.au/anz-guidelines/guideline-values/derive/cultural-values/principles

Moore, S. (2014). *Cultural competency in Central Australia* [Unpublished paper]. Northern Territory Department of Education.

Munns, G., & Woodward, H. (2006). Student engagement and student self-assessment: The REAL framework, assessment in education. *Principles, Policy & Practice, 13*(2), 193–213. https://doi.org/10.1080/09695940600703969

Nanni, G., & James, A. (2013). *Coranderrk: We will show the country*. Aboriginal Studies Press.

National Curriculum Board. (2008). *The shape of the national curriculum: A proposal for discussion*. Australian Curriculum, Assessment and Reporting Authority. Retrieved January 11, 2023, from https://docs.acara.edu.au/resources/The_Shape_of_the_National_Curriculum_paper.pdf

National Curriculum Board. (2009). *The shape of the Australian curriculum*. Australian Curriculum, Assessment and Reporting Authority. Retrieved January 11, 2023, from https://docs.acara.edu.au/resources/The_Shape_of_the_Australian_Curriculum_May_2009_file.pdf

National Institute for Direct Instruction. (2020). *Basic philosophy of Direct Instruction (DI)*. Retrieved January 11, 2023, from https://www.nifdi.org/what-is-di/basic-philosophy.html

National Native Title Tribunal. (2014). *Native title determination details: DCD2014/010 – Grandplover* [pseudonym].

Northern Territory Board of Studies. (2017). *Keeping Indigenous languages and culture strong: A plan for the teaching and learning of Indigenous languages and cultures in Northern Territory Schools.* Northern Territory Government. Retrieved January 11, 2023, from https://education.nt.gov.au/__data/assets/pdf_file/0012/413202/Policy_Keeping-Indigenous-Languages-and-Cultures-Strong-Document_web_updated.pdf

Northern Territory Department of Education. (2017a). *Guidelines for the implementation of Indigenous languages and cultures programs in schools.* Northern Territory Government. Retrieved January 11, 2023, from https://education.nt.gov.au/__data/assets/pdf_file/0004/471712/indigenous-languages-and-cultures-guidelines.pdf

Northern Territory Department of Education. (2017b). *NT Indigenous languages and cultures: First languages maintenance (L1M).* Northern Territory Government. Retrieved January 11, 2023, from https://education.nt.gov.au/__data/assets/pdf_file/0020/715232/ntilc-l1m-nov18-03.pdf

Northern Territory Department of Education. (2019). *Framework for inclusion 2019–29: Inclusive education for all children and students.* Retrieved January 11, 2023, from https://education.nt.gov.au/__data/assets/pdf_file/0010/662779/Framework-for-inclusion-brochure.pdf

Northern Territory Department of Education. (2021). *Indigenous education strategy employment pathways: Communication and problem solving: Numeracy – Area (Version 2.2).* Northern Territory Government.

Northern Territory Government. (2016). *School handbook.* Northern Territory Government.

Northern Territory Government. (2018). *Angarapa ALT [Aboriginal Land Trust] NT NRM [Northern Territory Natural Resource Management] report.* Retrieved July 22, 2019, from NT Infonet: http://www.infonet.org.au

Ober, R. (2009). Both-ways: Learning from yesterday, celebrating today, strengthening tomorrow. *Australian Journal of Indigenous Education, 38*(Supplement), 34–39. https://doi.org/10.1375/S1326011100000806

Ober, R., & Bat, M. (2007). Paper 1: Both-ways: The philosophy. *Ngoonjook: A Journal of Australian Indigenous Issues, 31*, 64–86. Retrieved January 11, 2023, from https://eprints.batchelor.edu.au/id/eprint/99/1/Ngoonjook31_Ober&Bat.pdf

Olsen, T. A. (2017). Privilege, decentring and the challenge of being (non-) Indigenous in the study of Indigenous issues. *The Australian Journal of Indigenous Education, 47*(2), 206–215. https://doi.org/10.1017/jie.2017.16

Perso, T., & Hayward, C. (2015). *Teaching Indigenous students: Cultural awareness and classroom strategies for improving learning outcomes.* Allen & Unwin.

Philippoff, P., & Cox, E. (n.d.). *Measuring abundance: Transects and quadrats. Concepts: estimation, percentage, error and zonation.* University of Hawaii. Retrieved January 10, 2023, from https://www.hawaii.edu/gk-12/opihi/classroom/measuring.pdf

Pledger, M. (2018). *Cultivating culturally responsive reform: The intersectionality of backgrounds and beliefs on culturally responsive teaching behavior* [Doctoral dissertation, University of California San Diego]. UC San Diego eScholarship. Retrieved January 11, 2023, from https://escholarship.org/uc/item/4m44543x

Ravitch, D. (2010). *The death and life of the great American school system: How testing and choice are undermining education*. Basic Books.

Reconciliation Australia. (2019, February 7). *RA IGA s 2018, category B finalist, Apmer Akely-akely for the SCFC Utopia project* [Video]. YouTube. Retrieved October 12, 2022, from https://www.youtube.com/watch?v=9k3qwfvgkZY

Reicher, S. D., Haslam, S. A., & Platow, M. J. (2018). Shared social identity in leadership. *Current Opinion in Psychology*, *23*, 129–133. https://doi.org/10.1016/j.copsyc.2018.08.006

Rigney, L. I., Garrett, R., Curry, M., & MacGill, M. (2020). Culturally responsive pedagogy and mathematics through creative and body-based learning: Urban Aboriginal schooling. *Education and Urban Society*, *52*(8), 1159–1180. https://doi.org/10.1177/0013124519896861

Riley, L., & Genner, M. (2011). Bemel-Gardoo: Embedding Aboriginal cultural content in the science and technology syllabus. In N. Purdie, G. Milgate, & H. R. Bell (Eds.), *Two way teaching and learning: Toward culturally reflective and relevant education* (pp. 35–47). Allen & Unwin.

Ryan, C. (2016, February 2). *Inclusion and diversity in education – the imperative to challenge*. School Education Gateway. Retrieved January 11, 2023, from https://www.schooleducationgateway.eu/en/pub/viewpoints/experts/inclusion_and_diversity_in_edu.htm

Sarra, C. (2005). *Strong and smart: Reinforcing Aboriginal perceptions of being Aboriginal at Cherbourg State School* [PhD thesis, Murdoch University]. Murdoch University Research Repository. Retrieved January 11, 2023, from http://researchrepository.murdoch.edu.au/id/eprint/1687

Sarra, C. (2011). *Strong and smart: Towards a pedagogy for emancipation: Education for first peoples*. Routledge.

Scherer, P., & Steinbring, H. (2006) Noticing children's learning processes – Teachers jointly reflect on their own classroom interaction for improving mathematics. *Journal of Mathematics Teacher Education*, *9*, 157–185. https://doi.org/10.1007/s10857-006-0004-7

Teacher Registration Board of the Northern Territory. (2019). *Code of ethics for Northern Territory teachers*. Northern Territory Government. Retrieved January 11, 2023, from https://www.trb.nt.gov.au/system/files/uploads/files/2019/CodeOfEthics.pdf

Turner, M. (2005). *Bush foods: Arrernte foods from Central Australia* (2nd ed.). IAD Press.

United Nations. (2007). *United Nations declaration on the rights of Indigenous peoples*. Australian Human Rights Commission. Retrieved January 11, 2023, from https://humanrights.gov.au/our-work/un-declaration-rights-indigenous-peoples-1

White, L., & Berta, L. (2019). *Review of education services in Homeland Learning Centres* [Unpublished draft report]. Northern Territory Government.

White, M., Albrecht, D., Duguid, A., Latz, P., & Hamilton, M. (2000). *Plant species and sites of botanical significance in the southern bioregions of the Northern Territory: Vol. 1. Significant vascular plants, Part 1: Species of significance*. Arid Lands Environment Centre.

Wilson, B. (2014). *A share in the future. Review of Indigenous education in the Northern Territory*. The Education Business. Retrieved January 11, 2023, from https://www.nt.gov.au/__data/assets/pdf_file/0020/229016/A-Share-in-the-Future-The-Review-of-Indigenous-Education-in-the-Northern-Territory.pdf

Woolley, T., Sivamalai, S., Ross, S., Duffy, G., & Miller, A. (2013). Indigenous perspectives on the desired attributes of medical graduates practising in remote communities: A northwest Queensland pilot study. *The Australian Journal of Rural Health*, *21*(2), 90–96. https://doi.org/10.1111/ajr.12018

Zimmerman, B. J. (1995). Self-efficacy and educational development. In A. Bandura (Ed.), *Self-efficacy in changing societies* (pp. 202–231). Cambridge University Press. https://doi.org/10.1017/CBO9780511527692.009

CHAPTER 2

Validating Indigenous Pedagogies in Mathematics Education

In Australia, like other colonised countries, secondary school mathematics education seldom connects the culture of the learner to mathematics learning in "pedagogically informed ways" (Madusise, 2015, p. 14). Despite this, pedagogically validating ways of teaching have been conceptualised by Australian Indigenous scholar Professor Lester-Irabinna Rigney and his colleagues (Rigney & Hattam, 2020). Rigney, who is from Narungga, Kaurna, and Ngarrindjeri nations in South Australia (Janke, 2019, p. 258), provides a framework for teaching that "pedagogically validate[s]" the Aboriginal learner's "cultural intelligences" (Rigney, 2019a, para. 12). He advocates for the teacher to share the power in the classroom with Aboriginal learners, their parents, and elders since

> children are more likely to meaningfully engage in learning when their identities belong, and their socio-cultural context is invoked. Therefore, educators must cultivate authentic relationships that position students, parents and Elders as co-designers toward effective and culturally responsive learning. (Rigney, 2019a, para. 12)

Furthermore, Rigney's colleague Professor Hattam asserts that:

> As soon as young people are given more control over what's going on, then that changes the nature of the relationship between the teacher and the student, and from our perspective increases engagement and increases student achievement. (Rigney & Hattam, 2020, 40:35)

1 Are First Nations Identities and Cultural Intelligences Invoked in Mathematics Learning?

Holding "affirming and anti-deficit views" of students and teaching "to and through (learners) cultural strengths, and intellectual capabilities" invokes a student's sense of belonging (Rigney, 2019a, paras. 6, 14). When students experience a "substantive sense of satisfaction with, and a psychological investment in, the classroom work being undertaken" they authentically engage in

 | DOI:10.1163/9789004682757_002

learning (Munns & Woodward, 2006, p. 195). Disengaging with curriculum, by contrast, can occur when students perceive that their knowledge and skills are not valued at school. The Australian Government recognises the risk of First Nations students disengaging entirely from education during the middle years – Years 6–9 (Education Council, 2019, p. 13). To address this risk, the Alice Springs (Mparntwe) Education Declaration (Education Council, 2019, p. 13) recommends all Australian schools focus their efforts on the "range of needs" of each learner to improve their "motivation and engagement" in middle years education. Meeting the needs of every learner and creating equity and excellence requires "adaptive expertise" and "purposeful collaboration" among educators (Halbert & Kaser, 2019, p. 22). Australian Indigenous students should have access to high-quality education while strengthening their sense of cultural identity (Education Council, 2019; Perso & Hayward, 2015). With suitable pedagogic adjustment, Indigenous middle years students can "embrace their cultures, languages and identities as Australia's First Nations peoples" while being provided with opportunities to access all levels of mathematics (Education Council, 2019, p. 16).

This research re-imagines measurement pedagogic design to incorporate the NTILC curriculum framework (see Chapter 1, Figure 6; NTDOE, 2017). This includes who speaks; ways of talking; building connections between home/community and school language and culture; discussion with local community on appropriate knowledge concepts relating to akatyerr for students to construct new measurement knowledge; and discussion on where the Indigenous knowledge relating to akatyerr can be shared. The inextricable relationship between Australian Indigenous people and the land is reflected in these words from Amelia Turner: "The land needs words, the land speaks for us and we use the language for this. Words make things happen – make us alive. Words come not only from our land but also from our ancestors" (House of Representatives Standing Committee on Aboriginal and Torres Strait Islander Affairs, 2012, p. 10).

1.1 *Communication of Mathematics*

The way mathematics knowledge is communicated in the classroom can be exclusionary. That is, there is a tendency for Australian secondary mathematics teachers to communicate knowledge in a one-way interaction communicated from the teacher to the student (Scherer & Steinbring, 2006). The interaction is limiting as it denies students opportunities to engage with deep content knowledge. Students need to develop skills that "support imagination, discovery, innovation, empathy" and create their own solutions to complex problems (Education Council, 2019, p. 15) rather than learning how to solve

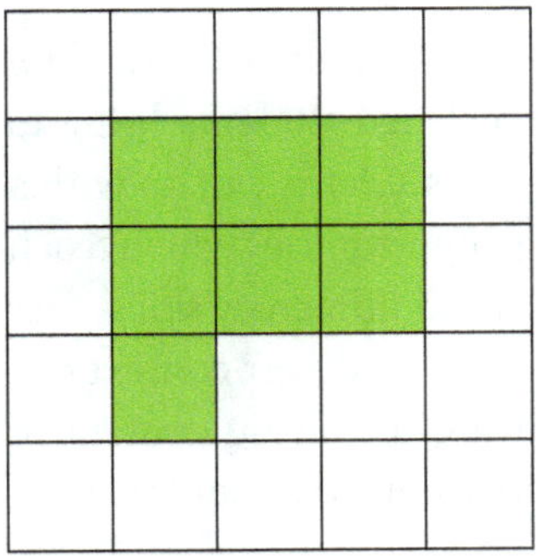

_________ Square units

FIGURE 7
Example question to demonstrate disciplinary thinking of area of the region in green
SOURCE: NORTHERN TERRITORY DEPARTMENT OF EDUCATION, 2021, P. 34

routine exercises demonstrated in a secondary mathematics textbook (Ewing, 2011). In Northern Territory remote schools, for instance, students are expected to solve area problems in a region or an array using a small number of squares such as the image shown in Figure 7 (NTDOE, 2021, pp. 34–37).

The Australian measurement curriculum, when communicated in this way, provides very limited academic challenge to remote Northern Territory Indigenous students. The Australian Government acknowledges that students flourish when schools "provide a culturally inclusive curriculum" delivered by "skilled teachers with high expectations" (Australian Government Department of Prime Minister and Cabinet, 2020, p. 58).

As stated earlier, students are considered at risk of disengaging entirely from middle years (Years 6–9) education (Education Council, 2019, p. 13) if they have inadequate opportunity to access age-appropriate achievement standards (see ACARA, 2014). Teaching and learning opportunities which simultaneously respond to the cognitive, emotional, and behavioural needs of the learner are likely to engage students (Munns & Woodward, 2006). Student engagement is elevated when learning involves "deep understanding and expertise" (high cognition), combined with a sense that students "genuinely value" what they are learning (high emotion) and are "actively participating" (high behaviour) in learning (Munns & Woodward, 2006, p. 194).

The Australian Curriculum framework allows teachers the flexibility to negotiate and select contexts for measurement teaching and learning (Perso & Hayward, 2015). Negotiating context is especially important for students living in remote Indigenous communities. Indigenous Australians living in remote contexts may have limited exposure to, and opportunity to quantify, particularly large cardinal numbers (Perso & Hayward, 2015). Additionally, if students are encouraged to work with a range of numbers (large whole numbers, ratio, percentage, fractions, decimals) they are more likely to think flexibly (Siemon et al., 2005). For remote teachers to implement the Australian Curriculum in ways that are inclusive of the mathematics knowledge students bring to school, it is important to first find out about what the students can do with a range of numbers. Teachers can then build learning opportunities which teach to and through students' interests, lived experiences and numeracy skills. Year 8 Australian mathematics students living in a very remote desert context can communicate their proficiency with "problem-solving" involving content in the measurement strand by "modelling practical situations" that, for instance, involve "areas of common shapes" (ACARA, 2020).

To design culturally inclusive curriculum with high expectations of students, it is important that mathematics assessments include a component of numeracy to investigate what skills students have. Numeracy should be considered a capability rather than a body of knowledge (Perso & Hayward, 2015). That is, students who can "choose the appropriate mathematics to use in particular situations and successfully apply the mathematics they know to this range of contexts" are considered numerate (Perso & Hayward, 2015, p. 103). Students' experience with computation of a range of numbers in different social contexts promotes numeracy. As Harris (1980, p. 15) concluded, it is the "social context" of a learning activity that influences "the degree of mathematical skill" demonstrated by students. Students depicted in Figure 8 are familiar

FIGURE 8 Quadrat survey design to solve problems involving areas of common shapes. Modelling a practical situation – akatyerr vegetation cover is surveyed along a 50-metre transect line to construct new mathematics knowledge
PHOTOGRAPH BY TIM WARE

with the social context when participating in school bush trips and identifying akatyerr and other plants. Students involved in the akatyerr survey had elected to participate in this bush trip. They had indicated their interest in furthering their knowledge about protecting akatyerr harvesting sites while participating in yarning circles before the site visit.

The following section describes how this study enabled Indigenous pedagogies to be validated in measurement learning.

2 A Framework for Validating Indigenous Pedagogy

Professor Lester-Irabinna Rigney's (1999) three Indigenist principles were used to guide the research: privileging Indigenous voice, resistance as the emancipatory imperative, and political integrity. Additionally, three Indigenous values identified in Kendall et al.'s (2011) cultural respect framework – respect, responsibility, and protection – were interconnected with Rigney's principles. Figure 9 depicts this interconnection. The central node, the cultural respect

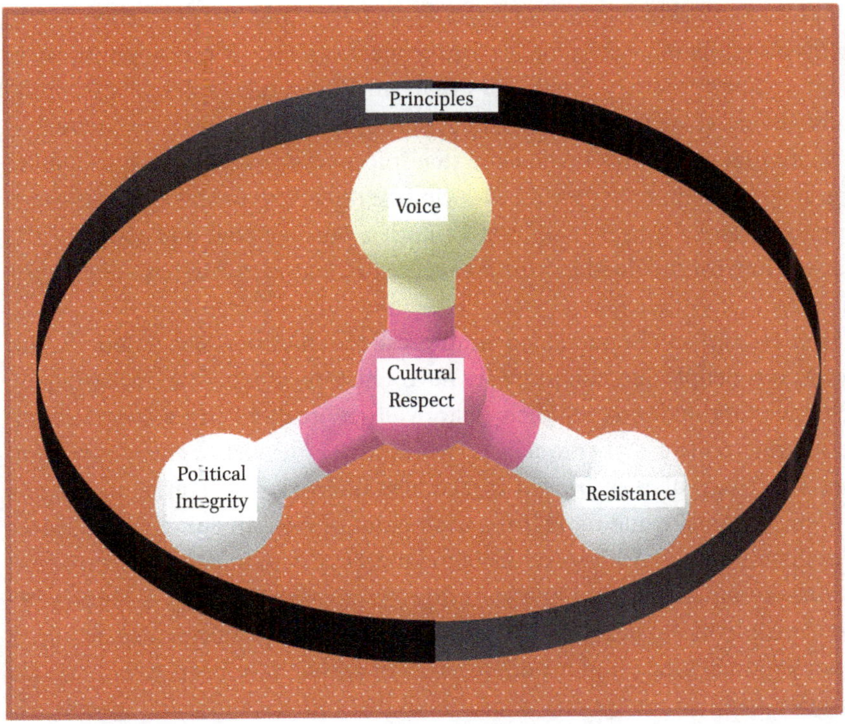

FIGURE 9 Principles of Indigenous research. The three principles – voice, political integrity, and resistance – are bound by a framework of cultural respect

framework (Kendall et al., 2011), was fundamental to the Indigenous research in my study. Rigney's (1999) three Indigenist research principles are represented in the three surrounding nodes: privileging Indigenous voice, resistance (as the emancipatory imperative), and political integrity. These were essential to the Indigenous research in my study. The figure is represented as a three-dimensional model to emphasise the essential connection binding each of Rigney's principles to the cultural respect framework.

Combined, the principles of Indigenous research ensured the study was conducted through respectful relationships and a responsibility to a collaborative pedagogic decision-making process, and that the retelling of information shared prioritised the protection of Indigenous cultural and intellectual property (ICIP). The adaptation of each principle to my study is addressed in the following subsections. Rigney's principle of privileging Indigenous voice is given priority as it primarily informed the study.

2.1 *Privileging Indigenous Voice*

This study drew on the historical traditions, lived experiences, and stories of Australian Indigenous students, assistant teachers, and community members, alongside the Australian non-Indigenous teacher's perspective, when teaching the Australian measurement curriculum. Therefore, it was appropriate to adopt Rigney's guiding principle of privileging Indigenous voice to inform the study. Giving voice to Indigenous people responds to the need for "increased participation and input" in dialogue that informs decisions directly related to their lives (Gough, 2008, p. 224). Drawing on the voices of the Utopia Eastern Anmatyerr and Alyawarr community stories, perspectives, and views of the akatyerr harvest provided important and valuable stories that have often gone unheard.

Smith (1999) and Rigney (1999) emphasise that past research practices with Indigenous people and communities have a history of exploitation, suspicion, misunderstanding, and prejudice. Historically, more dominant voices predominated, such as those of colonialists, with Indigenous voices given less attention. The views of Westernised researchers often skewed stories to address their own agenda when retold from their point of view. In doing so, the voices of Indigenous people were excluded and marginalised. In contemporary education settings such as in Canada, Indigenous academics give voice to Indigenous people by revealing the "wealth and richness of Indigenous languages, cosmology, teachings, and experiences, all of which have been systematically excluded from contemporary educational institutions and from Eurocentric knowledge systems" (Battiste, 2005, p. 1). However, it is not possible to oppose the "denial of Indigenous voice" in decision-making processes in education

(McConaghy, 2000, p. 243) by simply advocating to include Indigenous voice. Perceiving the political process of inclusion of Indigenous voice as negating exclusion does not alter the issues that "result in oppressions and disadvantages in contemporary times" (McConaghy, 2000, p. 238).

Arbon (2008, pp. 110–111) identified that research in Australian Indigenous contexts is problematic since it is difficult for an outsider to determine whose voices are to be privileged in communities. Speaking about specific topics should be determined by relatedness among the members of the community or kinship and not by an outsider. I was an outsider to the community but held a teaching position in the community; therefore, I became an insider/outsider. Although there is much scholarly debate about the validity of insider/outsider research (see, for example, Innes, 2009; Martin, 2008; Smith, 1999), my position as a teacher within the Utopia Eastern Anmatyerr and Alyawarr community allowed me the opportunity to work closely with assistant teachers and other community members. Community involvement in the research process allowed for the development of meaningful Utopia Eastern Anmatyerr and Alyawarr perspectives, established positive and respectful relationships, and gave authenticity to local perspectives and voices. With the responsibility I held in my position as an insider/outsider, it was important that I acted with an awareness of, and operated in fairness to, the relatedness among the Utopia Eastern Anmatyerr and Alyawarr community members and their voices.

I sought members of the Utopia Eastern Anmatyerr and Alyawarr community to represent voice. The elements that determined appropriate people and their responsibilities to the akatyerr harvest site investigated are elaborated in Chapter 4. Without awareness of the complexities of the relatedness among community members, I may have marginalised (Arbon, 2008) the voices of other Utopia Eastern Anmatyerr and Alyawarr members who had a particular responsibility for the akatyerr cultural knowledge shared in the study.

The cultural advice given by the eight adult participants from the Utopia Eastern Anmatyerr and Alyawarr community (two assistant teachers and six community members) and 12 middle years students was sought to ensure appropriate guidance for the treatment of akatyerr cultural knowledge in the study. This required that I really examine my own beliefs about the learner, the subject knowledge, my teaching, and curriculum (Rigney & Hattam, 2020). As stated by Professor Rigney in his discussion with Professor Hattam on teaching to and through Aboriginal learners' cultural strengths:

> Unless you're reassessing as educators our view of the learner and a view of their life worlds, the view of ourselves as educators, the view of what learning is, the view of knowledge, what knowledge is, whose knowledge

> is privileged in my classroom, the voice that we give to students and how we play that out and allow some students to speak over others and other voices to be silent over others ... Once we start to really ... examine these, and how we see these as the teacher, gets us to a far more comprehensive, enriching relationship with our students. (Rigney & Hattam, 2020, 41:18–42:13)

2.1.1 Privileging Utopia Eastern Anmatyerr and Alyawarr Voice in the Akatyerr Study

I created more opportunities for authentic relationships by privileging the voices of Utopia Eastern Anmatyerr and Alyawarr speakers. I reassessed the view I had of whose knowledge is privileged in the classroom to inform ways to adjust the measurement (length and area) curriculum and pedagogic practices. Mathematics education based on more authentic relationships resulted in more inclusive education practices. I envisioned ways to enable my students to move flexibly between the familiar concepts of akatyerr harvest knowledge and the unfamiliar Western concepts of measurement, such as conducting a quadrat survey. The study recorded the adjustments that enabled "direct dialogue" (McConaghy, 2000, p. 15) with Utopia Eastern Anmatyerr and Alyawarr students, assistant teachers, and community members to contextualise the measurement curriculum to sustainable wild akatyerr harvest.

As the research was conducted at a culturally significant akatyerr growing site within a broader region (Grandplover – a pseudonym) that had recently been granted non-exclusive native title rights (National Native Title Tribunal, 2014), it was essential to contact and refer to the appropriate cultural authority and voices responsible for this region. The native title rights of the area surrounding the research site co-exist with perpetual pastoral leases. This means the region continues to be managed by customary law and the *Native Title Act 1993* (Australian Government, 2021). The local land council overseeing permit applications was the most appropriate entity to contact first, for me to be granted permission to conduct the study on an akatyerr growing site. Access was granted because I was also a member of the school community and had a responsibility as an employed secondary teacher to deliver contextualised learning on country at a range of sites within this jurisdiction.

An informed-consent meeting was held to explain the project. Authority to conduct the study was obtained from the local Aboriginal Board members who attended this meeting. Cultural authority from parents, assistant teachers, and the wider school community was also sought at this meeting so that the voices of those attending could direct the appropriateness of conducting the research at an akatyerr harvest site. The informed-consent meeting explained the concept of free prior and informed consent in Utopia Eastern Anmatyerr and Alyawarr language, and members attending the informed-consent presentation had 5 weeks to consider whether to "opt in".

Rigney considers it appropriate that Indigenous Australians "give voice" to other Indigenous people (1999, p. 118); therefore, the appropriate Indigenous cultural authority was sought as already stated. The Utopia Eastern Anmatyerr and Alyawarr assistant teachers and community members who decided to participate, or opt in, self-identified as appropriate cultural authorities to inform the ways to conduct proceedings during yarning circles on akatyerr. Later, additional Utopia Eastern Anmatyerr and Alyawarr participants known through kinship considerations to have significant cultural obligation to the chosen akatyerr harvest site joined the research effort. That is, adult participants who had a responsibility as akatyerr harvesters (women) or traditional land managers (young men or traditional owners) identified their interest in joining the yarning circles at the chosen akatyerr site.

Privileging Indigenous voices allowed for Indigenous adults and students to "be respectful of each other's cultural traditions" (Rigney, 1999, p. 118). It allowed Indigenous assistant teachers to give voice to Indigenous students' interests and experiences. Yarning circles (Woolley et al., 2013) created the space for negotiation of appropriate akatyerr knowledge concepts to respond to and include the "social cultural context of the learner" (Davis & Grose, 2008, p. 11). This space enabled all participants a share in determining what the "ways of proceeding look like" when adjusting measurement curriculum and pedagogic practice to reflect students' interests and experiences (Scherer & Steinbring, 2006, p. 174). Further explanation of appropriate ways of acting is provided in Chapter 4.

The "underlying idea[s]" that emerged from a thematic analysis of the audio- and video-recorded dialogue on knowledge concepts interrelated to akatyerr were translated and transcribed "for publication purposes" (Janke & Cubillo, 2019, p. 3). Thematic analysis is discussed in Chapter 4. The next section discusses Rigney's (1997, 1999) Indigenist principle of political integrity.

2.2 *Political Integrity*

Rigney (1999) does not argue that research conducted by non-Indigenous people should not continue. Neither does he argue that such research cannot serve to inform the struggles of Indigenous Australians as they seek genuine self-determination. Recognising the Indigenous struggle for self-determination is an integral part of political integrity. The need for Indigenous Australians to determine their political agenda for liberation can be achieved when it derives from Indigenous people in their communities with their people. That is, the principle both recognises the survival of Indigenous Australians from their continued resistance to genocide and attempts to reveal and seek healing from the issues that have stemmed from the long history of oppression (Rigney, 1997). Similar to Rigney's research efforts to advocate for Indigenous Australians (1997), the Canadian Indigenist research effort is primarily about informing the Indigenous struggle for self-determination so Indigenous students can be engaged in education that is "not assimilative or racially defined" (Battiste, 2005, p. 5).

The historical situation of denying Indigenous people self-determination and self-representation in education is shaped by the "cultural and racist social engineering theories" that underpinned the Australian Government's creation of the assimilation policy (Rigney, 1997, p. 1). The power to subordinate Indigenous people was enacted through this policy. It created deep-rooted racism in Australian society, which persists through "silent assumptions" associated with belonging to an inferior or superior race (Langton, 2013, p. 6). In remote Australia, entrenched racism has become so normalised that it is essentially "invisible" (Guenther et al., 2017, p. 200).

Non-Indigenous Australians' perceptions of Indigenous Australians continue to be influenced by these assimilation policies and Darwinian concepts of race (Rigney, 1997, p. 1). A pervading perception that Indigenous Australians belonged to an inferior race has been legitimised by this racist ideology (Arbon, 2008). Rigney (1997) debates anticolonial cultural criticism of Indigenous Australian representations because they continue to be subjected to subordination from exclusionary educational processes in Australian society (Rigney, 1997). The social and cultural impacts from these policies are pervasive, as argued by Maynard:

> The aftermath of 1788 witnessed a long period where historical erasure and fracture clearly targeted Aboriginal people. We were driven and encouraged into a state of forgetting and detachment from our past. The methods used by government to ensure the genuine absence of an Aboriginal place in the landscape were dispossession, segregation, assimilation and cultural/historical deprivation. Combined, these policies proved extremely destructive and today, Aboriginal Australians still carry the psychological scars together with sad and scattered memories of terrible experiences in their personal pasts. (Maynard, 2007, p. 118)

The continuance of oppression confronts and inhibits Indigenous Australians and their access to education. It continues to be played out through policy at the federal and state levels of government. Rigney (1999) argues that there are ways to "decolonise epistemologies" (1999, p. 109) so that the construction of new knowledge liberates Indigenous people. The racialised epistemology of positivist traditions, well documented in scientific research but also identified as persistent in Indigenous education, continues to oppress Indigenous people as they participate in the "very social formations" (racialised research practices) responsible for their oppression (Rigney, 1999, p. 113). Moreover, it is possible that "the methods of constructing new knowledge" (Rigney, 1999, p. 111) conducted without appropriate "strategies to decolonise epistemologies and create new ones" (Rigney, 1999, p. 110) could continue to colonise Indigenous Australians. This continuance works to thwart the recognition of self-determination for Indigenous Australians.

Recognition involves revealing and responding to injustices from opposing positions of authority, for instance, from the coloniser and colonised power relations (Lyotard, 1984/1986, 1983/1988). That is, as French philosopher Jean-François Lyotard revealed, injustices in Australia have occurred where differences exist, such as between native title law (Australian Government, 2021) and traditional lore, which "cannot be equitably resolved for lack of a rule of judgement applicable to both" (Woodward, n.d., "The Differend" section, para. 1). Recognition in this study involved "the appreciation, support, and defence of cultural differences" among participants (Buckley & Quaye, 2016, p. 1119). Recognition for self-determination of Indigenous Australians, in this study, involved responding to past racialised injustices by hearing from "narratives, still largely unwritten" (Peters, 1995, p. 395). Further, the study responded to injustices by generating a theory that sought to "critique and dethrone existing" deficit assumptions which legitimise limiting teaching and learning practices in Indigenous classrooms (Peters, 1995, p. 395).

Moreover, "Indigenous people have a fundamental right to expect research and epistemologies to address the issues and racialising practices" (Rigney, 1999, p. 114) that occur "within the methods of constructing new knowledge" (Rigney, 1999, p. 111). This is pertinent since research practices grounded in the concept of race deny the value of the "minds, intellects, knowledges, histories and experiences" (Rigney, 1999, p. 113) of Indigenous Australians.

The next subsection further discusses the application of Rigney's Indigenist principle of political integrity in this study.

2.2.1 Political Integrity: Utopia Eastern Anmatyerr and Alyawarr Community

Political integrity is described as "research that seeks to uncover and protest the continuing forms of oppression" confronting Indigenous Australians"

(Rigney, 1999, p. 117). It seeks to support their struggles and carve out a way of being in Australia where there can be healing from historical oppression and cultural freedom in the present and future.

This principle assisted me to define the purpose of analysing the data gathered and to observe the integrity of Indigenous people (the Utopia Eastern Anmatyerr and Alyawarr community) through their involvement or lack of representation within the education of their young people. It allowed me to identify actions prescribed to encourage community engagement in schools. With partnerships being a key strategy in government policies (Education Council, 2019; NTBOS, 2017), this interaction or lack thereof was important in the study. My study recognised the rights of Indigenous educators for self-determination. This is particularly relevant when making decisions about the way Indigenous knowledge is treated, when enacting NTILC curriculum, and when making links to the Australian measurement curriculum. This is pertinent since "Indigenous educators emphasise the importance of Indigenous ownership and control over Indigenous knowledge, curriculum and pedagogy, in relationships of equality with non-Indigenous knowledge, curriculum and pedagogy" (NTBOS, 2017, p. 52).

The principle of political integrity was necessary to create a space for direct dialogue with assistant teacher participants (and parents/grandparents) on ways to contextualise pedagogy and measurement learning. Aboriginal parents "want their kids to improve in their leaning but also to maintain their identities and their sense of wellbeing and belonging" (Rigney & Hattam, 2020, 2:44–2:53). However, the use of a quadrat survey framed within a positivist tradition of scientific research (as previously discussed) could be seen as "denying the value and authenticity" of the assistant teachers' voices and knowledge (Rigney, 1999, p. 113). The care taken to include culturally grounded knowledge in the yarning circles facilitated students to cross over or traverse between the cultural authority of, for instance, a student's grandparent (also assistant teacher) and the cultural authority represented in Western epistemologies through the use of the scientific survey tool (the quadrat) in the akatyerr investigation. That is, efforts which bring "Indigenous mathematizing into school mathematics … in a non-tokenistic way" facilitate a "cross cultural" understanding of mathematics curriculum which "feels more familiar" to students due to it "having more human dimensions" (Aikenhead, 2017, p. 571; Aikenhead & Michell, 2011). The human dimensions are depicted in Figure 10 and in the following retelling of a conversation that occurred in this yarning circle.

The student's voice had trailed off and was unheard the first time she asked. A cacophony of laughter filled the air. The teacher and student laughed together. Then, the teacher gently spoke and prompted the student. Reminded

FIGURE 10 The human dimension to mathematics learning. This image depicts the interactions between community members, the teacher, and students in a familiar social context, the yarning circle
PHOTOGRAPH BY TIM WARE

to speak again, after laughing off her shame, the student inquired about her interest in the following dialogue:

> Teacher: Go again.
> Student: Should we burn this land here?
> Land manager/ranger: Yeah, a little bit.
> (Yarning circle BB2, June 13, 2019)

Speaking in a familiar social context (yarning circle) was critical for this student's voice to be heard. The student was shy, or nyerralth (shamed), at first (Green et al., 2019, p. 397). However, supported by the community members, teacher, and peers, she was able to quickly overcome her shame and continue her investigation. The dialogue revealed to the student the areas within the site that were appropriate to burn.

A big pedagogical idea in measurement learning – that is, turning the continuous into the discrete – could sit alongside the NTILC curriculum framework, which aims to strengthen home and community (first) language and cultures (NTDOE, 2017). This aligns with Kaupapa Māori theory and its uses in decolonising methodologies (Smith, 1999; see also R. Bishop et al., 2007) to "reduce the epistemological difference between home and school" (Rigney, 2017, 43:50).

The pedagogical practices of actively listening and responding to the right configurations of people, place, and time (Christie et al., 2010) within the yarning circle and quadrat survey supported student agency over concepts of measurement (length and area). Moreover, these practices permitted students to access and maintain knowledge concepts interrelated to akatyerr, while constructing multiple representations of their measurement learning. If new knowledge is seen "not [as] a learned reflection of the world" but rather knowledge that "shapes the world in particular ways, for particular interests" (Waldby, 1995, p. 15), then the use of the quadrat can sit alongside the Utopia Eastern Anmatyerr and Alyawarr perspectives on akatyerr land management, as the interests are not in opposition. Both perspectives – here, Indigenous and Western – assess the management of land to promote sustainable wild harvest of akatyerr. In doing so the learning promoted "high intellectual challenge" and connected to the "life world of students", which is critical to "captur[ing] most of the intelligences that's brought from home to class and building on those" (Rigney & Hattam, 2020, 12:31).

Figure 11 depicts the interactions between a familiar community member, Jimmy Navie (Wanta Aboriginal Corporation), and two students while they were constructing new mathematics knowledge using the quadrat survey.

The next section discusses Rigney's Indigenist principle of resistance as the emancipatory imperative.

FIGURE 11 Jimmy Navie (right) from Wanta Aboriginal Corporation helps two students construct new mathematics knowledge about sustainable wild akatyerr harvest using the quadrat survey
PHOTOGRAPH BY TIM WARE

2.3 *Resistance as the Emancipatory Imperative*

Rigney (1999) recognised the need for analysing the emancipatory potential in Indigenous education. He describes this principle as

> research that seeks to uncover and protest the continuing forms of oppression confronting Indigenous Australians. Moreover, it is research that attempts to support the personal, community, cultural, and political struggles of Indigenous Australians to carve out a way of being for ourselves in Australia in which there can be healing from past oppressions and cultural freedom in the future. (p. 117)

The underlying philosophical assumptions that I held when contextualising measurement curriculum needed to be critiqued in the study. As a non-Indigenous researcher, it was necessary to consider the research problem and research questions to address how and to what extent, through the study, I was enacting issues of power and dominance over the participants and thus contributing to the maintenance of these issues throughout the study, including in the classroom and within the community. I also needed to understand how to identify how power and dominance occurred and how it was resisted by participants.

Power and dominance can be maintained in a number of ways. For example, new knowledge on the "inclusion of Indigenous perspectives and voices" if constructed without a "critical analysis of inclusionary politics" (McConaghy, 2000, p. 246) may contribute to continuing to legitimise "racist oppression in research" (Rigney, 1999, p. 110). Uncritically accepting the inclusion of Indigenous perspectives and voice as a political process that inverts "colonial power relations" (McConaghy, 2000, p. 239) can work to maintain dominance. Further, subtle forms of oppression may remain unseen since the racialising practices of the coloniser are so deeply embedded in social structures that they appear the norm (Guenther et al., 2017; Rigney, 1999; Scheurich & Young, 1997). "Aboriginal concepts, values & practices ... must be made normative to enable Aboriginal learner success" (Rigney, 2019b, slide 38).

The resistance principle assisted me in defining my purpose in the study and how it was conducted. It assisted me with identifying and critiquing the actions and practices I was prescribing because they may have had a limiting effect. For instance, I critiqued my experience of change when adapting curriculum and pedagogic practices in measurement to be inclusive of the local Indigenous culture, beliefs, and experiences in relation to akatyerr harvest. Further, I recognised that because my own "social structures and beliefs are socially constructed" they were able to be "transformed through [my] social action" (Ewert, 1991, p. 355). The actions I initiated that promoted direct dialogue and the interactions among the assistant teacher/teacher, students, and disciplinary knowledge are further developed in Chapter 7. But to briefly explain here, my perspective on how to understand the adjustments to curriculum and pedagogic practices that enabled contextualisation of measurement learning involved detailed discussions with a critical friend (who is a linguist) and key adult participants to ensure the information shared was accurately and respectfully represented. I needed to ensure that I was informed by the appropriate cultural authority while I was privileging student voice to enable access to and construction of new measurement knowledge.

Equity in education in this study is defined as the process of co-constructing socially just curriculum and pedagogic practices that are more aligned with students' real-life experiences (see, for example, Applebaum, 2010; Baggett et al., 2020; Carter Andrews et al., 2016). Inequitable education, by contrast, leads to significant and persistent differences in achievement between students from dominant contexts and those who are marginalised and from culturally, linguistically, and economically diverse contexts. I critically reflected on inequity in education by reviewing my social location (race, class, gender) and the positioning of this location (as privileged) within the education system I was employed in. I examined issues of power by reflecting on a mismatch in social location between myself and the students and community I served. I became conscious of the inequities in the education system (Carter Andrews, 2014) through an awareness of my own power and dominance as a teacher in the school community.

This principle, resistance, challenges the traditional view of Indigenous people as victims of oppression and in need of charity (Foley, 2003). Foley (2003) argues that resistance in Indigenist research is "research undertaken as part of Indigenous Australia's struggle for recognition and self-determination, to research the survival and celebration of Indigenous Australia's resistance struggle, to uncover and stop the continuing forms of oppression against Indigenous Australia" (p. 48). Rigney and Hattam (2020) suggest "an activist orientation for learning" where students "as inquirers" learn how to "try to solve key issues that impact their lives" (26:35–26:40). This foregrounds students' life worlds and empowers students to produce their own knowledge to improve the lives of their families, elders, and communities. This can occur, for instance, by building a student's sense of "critical consciousness" (Ladson-Billings, 1995). Reflexivity or critical consciousness (see Section 4.1) "is equally important for teachers" (Morrison et al., 2019, p. 30). The resistance principle assisted me to critically reflect on how I engaged with Indigenous people and their ICIP relating to akatyerr (Janke & Cubillo, 2019). ICIP relating to akatyerr refers to "all the rights that Indigenous people have, and want to have, to protect their traditional acts and culture" and includes "knowledge of plants" (Janke & Cubillo, 2019, p. 1) such as akatyerr.

To "ensure that Indigenous people [were] engaged in a professional and respectful manner" (Janke & Cubillo, 2019, p. 2), I adopted the protocols developed by Terri Janke and Company (10 True Tracks principles; Janke, 2019) and implemented these into the akatyerr study. Terri Janke and Company is an Indigenous law firm that delivered a professional development workshop I attended in April 2019 overviewing the 10 True Tracks principles. The purpose of engaging the legal services of this firm was to "cover ICIP rights that are not recognised at law" in relation to "underlying idea[s] or information" that were told by the Indigenous assistant teacher participants to me and the student participants in yarning circles (Janke & Cubillo, 2019, p. 2). Janke and Cubillo are the names of the lawyers with the firm and the co-authors of the ICIP statement that I referred to during the study.

The ICIP strategy implemented included "the use of contracts" to obtain "free prior and informed consent" that was "continuously being engaged" between me and "relevant Indigenous people in authority" (Janke & Cubillo, 2019, pp. 2–4). This enabled me to protect the "secrecy and privacy" of information revealed in dialogue by the process of "checking the accuracy of typed text (transcribed)" and deciding if "typed text" representing [adult or student participants'] ideas "should be kept, changed or taken out of the final report" (Janke & Cubillo, 2019, pp. 2–4). The principle of resistance provided a means to address the engagement or lack thereof by Indigenous students within the decision-making process in education and their learning. The next section applies this principle to the community.

2.3.1 Resistance as the Emancipatory Imperative: Two-Way Dialogue on Akatyerr

I recognised the importance of this principle when exploring ways Utopia Eastern Anmatyerr and Alyawarr participants could inform and guide the opportunities for measurement learning to be contextualised to akatyerr knowledge concepts. That is, the right to self-determination for Utopia Eastern Anmatyerr and Alyawarr participants when influencing "what students learn, how they learn it and the way the school itself reflects the community's cultural values" (NTDOE, 2018, para. 6). The study afforded me opportunities to critique my taken-for-granted or legitimised Western cultural values in my teaching practices. I critically reflected on the existing pedagogical practices I used to teach measurement to a cohort of female middle years students (see Chapter 6). I examined how these contributed to "differential access to forms of knowledge"

(Smyth & Wrigley, 2013, p. 24). Reflecting over the recordings captured in yarning circles and my field notes provided me with opportunities to adjust my practices in order to promote access to learning about measurement when contextualised to akatyerr during the planned site visit.

The study highlighted the emancipatory potential when permitting students to influence adjustments made to the ways they learn. Critiquing my legitimised practices in my measurement teaching (see Chapter 7) was a pre-condition for the research effort to include the students' interests. Student agency was promoted by allowing students to learn about measurement attributes such as length and area, through contextualisation to their real-world experiences and interests in knowledge concepts interrelated to akatyerr.

Together the critical friend, assistant teachers, community member participants, and I critically reflected on the emancipatory potential of the students' akatyerr interests to inform the planned measurement learning at the akatyerr site. Listening to and being guided by Utopia Eastern Anmatyerr and Alyawarr adult participants was imperative in deciding the appropriateness of sharing the two-way dialogue on students' interests in akatyerr at the site and with a wider audience. That is, the adult Utopia Eastern Anmatyerr and Alyawarr participants assessed the emancipatory potential of the two-way dialogue exchanged by students in yarning circles (in written form) and decided the interests that were most suitable to investigate. This critical reflection on the students' akatyerr interests by Utopia Eastern Anmatyerr and Alyawarr adults before and during the visit to the harvest site provided students with access to two-way dialogue on akatyerr knowledge concepts of interest to them while maintaining their cultural obligations encoded through kinship.

Rigney (1999) provided the space to counter colonialism at its sites of oppression and power, the place where knowledge is constructed. The social construction of knowledge is what "shapes the world in particular ways and for particular interests" (Waldby, 1995, p. 15). The research, through the merging of Western and Utopia Eastern Anmatyerr and Alyawarr epistemologies and interests in akatyerr harvest knowledge concepts, represents the "aspirations of Indigenous Australians" (Rigney, 1999, p. 114). In reflecting the aspirations of Utopia Eastern Anmatyerr and Alyawarr people for "intergenerational knowledge transfer" of akatyerr harvest knowledge, and a Western pedagogic big idea of turning the continuous into the discrete using measurement of length and area, the research represented a merging of epistemologies (CLC, 2017, p. 73). The blended epistemologies helped to "reaffirm" Utopia Eastern Anmatyerr and Alyawarr "knowledges and cultures" and in so doing "strengthen the struggle for emancipation and liberation from oppression" (Rigney, 1999, p. 114). The appropriate ways new knowledge was to be constructed were determined among the Indigenous students, assistant teachers, community members, and researcher involved in the yarning circles prior to the akatyerr site visit. Considerations regarding which (adult) Utopia Eastern Anmatyerr and Alyawarr participants were able to speak about akatyerr knowledge concepts and decide a suitable site to conduct the examination of knowledge concepts relating to akatyerr were determined in initial yarning circles. Also, how students should act, including ways of speaking and listening (NTDOE, 2017), was considered by the adult participants in yarning circles prior to and during the akatyerr site visit.

In my study, new knowledge emerged by considering which akatyerr knowledge concepts were appropriate and how and where Utopia Eastern Anmatyerr and Alyawarr participants could contextualise these concepts with concepts of measurement. For instance, new ways for teaching and learning about concepts of measurement (such as length and area) developed by connecting (merging) students' lived experiences of akatyerr harvesting with Western pedagogic concepts such as isolating a unit of measure. Merging also occurred at the significant akatyerr patch as both cultural protocols and Western scientific methods were enacted or adhered to while the teaching and learning experiences occurred. That is, Utopia Eastern Anmatyerr and Alyawarr participants shared the responsibility (with the non-Indigenous educator) for conducting the quadrat study of akatyerr and yarning about how to understand and assess the sustainable management of this important plant food. Prioritising the voice of Indigenous participants enabled new knowledge to emerge about how two-way dialogue progressed measurement learning that occurred at a significant place where akatyerr was known to grow.

3 Cultural Respect Framework

The research was informed by complementary theoretical frameworks: an Indigenist research framework (Rigney, 1997, 1999) and a cultural respect framework (Kendall et al., 2011). Indigenist research frameworks like Rigney's (1997, 1999) emphasise the need for "researchers to adopt new ways of seeing that respect local Indigenous ways of knowing" (Kendall et al., 2011, p. 1719). First Nations peoples across the world have been repeatedly researched with "little respect given to culturally appropriate methods of engagement" (Brereton et al., 2014, p. 1183). To ensure that I participated in respectful ways, I was guided by elements of the cultural respect framework, in particular responsibility for and protection of ICIP (Janke & Cubillo, 2019; Kendall et al., 2011).

3.1 *Respectful Social Relationships*

The Utopia Eastern Anmatyerr and Alyawarr community and I, as a secondary teacher, built social relationships two years prior to the current study. We took the time "to talk and share space" in the classroom and community (Davis & Grose, 2008, p. 9). The social relationships between Indigenous assistant teachers and the non-Indigenous researcher were integral to the way we interacted and were understood. During the akatyerr study, the use of Utopia Eastern Anmatyerr and Alyawarr language and culture content in the measurement teaching and learning program provided a mechanism for Indigenous assistant teachers to "take on leadership roles" (Douglas, 2011, p. 24). For instance, I deferred to the Indigenous assistant teachers as experts on Utopia Eastern Anmatyerr and Alyawarr language, culture, and akatyerr measurement, and the Indigenous assistant teachers deferred to me as an expert on the contextualisation of measurement in the study design. This assisted in continuing to develop trust and respect between us (Douglas, 2011).

Respectful relationships enabled participants to share and combine ideas. For instance, we considered what would be a suitable blend of Indigenous and Western knowledge concepts to contextualise the measurement curriculum and pedagogical practices at the akatyerr site. While planning the akatyerr site visit, a Utopia Eastern Anmatyerr and Alyawarr assistant teacher shared an estimation of the capacity and mass of dried fruit likely to be found within a 100 cm × 100 cm area at the site. I incorporated this measurement in the Goompi pedagogical cycle (see Chapter 5) as the standard against which students' measures would be assessed. The ideas shared were built into the pedagogic design. Interaction patterns demonstrating that ideas were "valued, respected and acted upon" enabled a deeper understanding of sharing responsibility for developing curriculum and pedagogic practice (Douglas, 2011, p. 24).

3.2 *Responsibility in Pedagogic Decision-Making Process*

The study sought to embed a cultural respect framework into research practices to inform ways of proceeding from "culturally grounded knowledge" (Kendall et al., 2011, p. 1725). This entailed participants referring to an akatyerr concept map (see Figure 12) to guide discussions and work through suitable knowledge concepts interrelated to akatyerr to contextualise the measurement learning, together. During yarning circles, the use of the concept map provided a safe space to identify students' interests relating to akatyerr "without interruption" (Mills et al., 2013, p. 288), since the person holding the concept map was the only person permitted to articulate their views. The practice of using a yarning circle was established so that each contribution made by participants throughout the discussion was respected. I facilitated a discussion among student peers by employing the concept map as a focal point that acted like a

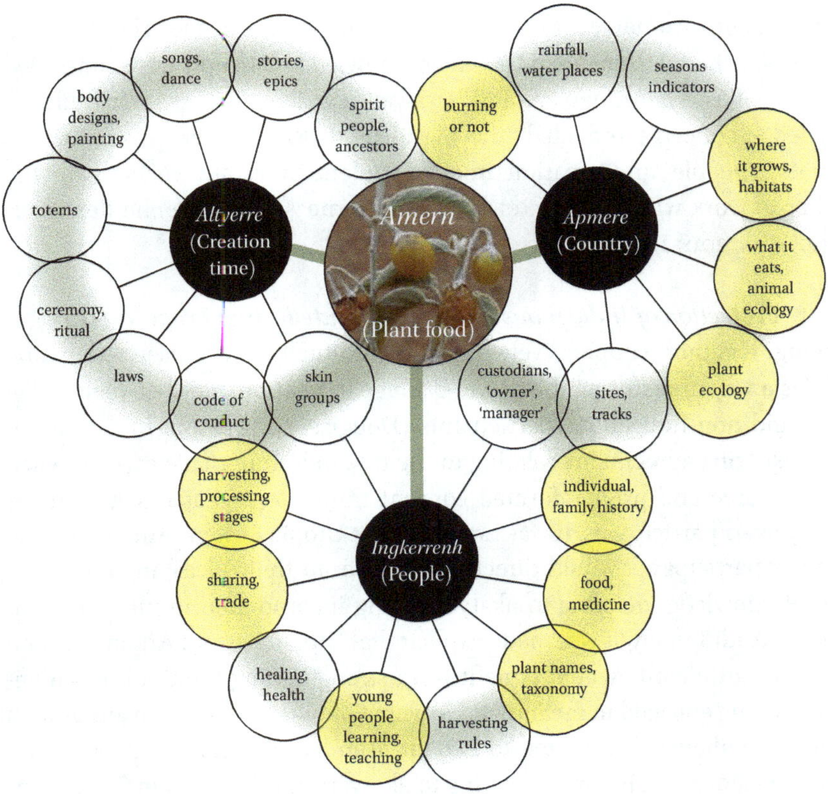

FIGURE 12 Knowledge concepts interrelated to akatyerr (adapted from Dobson et al., 2008, Merne Altyerre-ipenhe Reference Group et al., 2011, and Turner, 2005)

"message stick" (Mills et al., 2013, p. 288). That is, the students each had a copy of the concept map, a series of interconnected knowledge concepts known to be related to the akatyerr plant. Individuals chose which of the concepts they were interested in learning about and wrote these on the concept map. While writing their responses, what they said was captured without interruption on each participant's copy of the concept map. Each participant's contribution was respected whether they chose to participate or not. Participants in the yarning circle could choose to learn by "watching and listening, waiting and then acting" (Ungunmerr-Baumann, 2002, as cited in Mills et al., 2013, p. 288). That is, any participant could choose to articulate ideas or could also choose only to listen (Mills et al., 2013, p. 288). This was important as yarning circle dialogue facilitated intergenerational transfer of knowledge concepts interrelated to akatyerr such as knowledge of rainfall, plant names, and burning and harvesting the land.

Responsibility for pedagogic decision-making regarding the appropriate knowledge concepts in Figure 12 for contextualisation were shared between

Utopia Eastern Anmatyerr and Alyawarr participants and me. The decision-making was shared to ensure the akatyerr knowledge concepts were controlled by Utopia Eastern Anmatyerr and Alyawarr participants. Akatyerr cultural knowledge concepts provided "the big ideas" (NTBOS, 2017, p. 28) such as "country", "people", and "creation time" for students, assistant teachers, and the teacher to work with (Dobson et al., 2008; Merne Altyerre-ipenhe Reference Group et al., 2011; Turner, 2005).

3.3 *Protection of Indigenous Cultural and Intellectual Property* (ICIP)

Working together cooperatively towards "mutually recognised education goals" is a way of restoring the power balance in relationships between Indigenous and non-Indigenous participants (Douglas, 2011, p. 25). In this study, benefit-sharing agreements relating to the research were made explicit when obtaining free and prior informed consent from Utopia Eastern Anmatyerr and Alyawarr participants. It was expected that Utopia Eastern Anmatyerr and Alyawarr participants would directly benefit from their ideas, interests, and cultural knowledge relating to akatyerr being incorporated in measurement learning. Additionally, it was made explicit that Utopia Eastern Anmatyerr and Alyawarr participants would benefit from strengthening their cultural identity and being engaged in meaningful learning of measurement. Together with the wider audience, I expected to benefit from learning how to include cultural knowledge and interests relating to akatyerr in middle years (Years 8–9) measurement curriculum and pedagogic practices. These expectations were explicitly stated and explained in the informed-consent processes, which are elaborated in Chapter 6.

4 Critical Self-Reflection

4.1 *View of the Teacher*

Critical self-reflection on inequity in education is important professional work for teachers. The social location (race, class, gender) of the teacher and the positioning of this location (as privileged or underprivileged) within the education system helps teachers to identify and examine issues of power. Teachers, by reflecting on a mismatch in social location between themselves and the students they serve, become conscious of the inequities in the education system (Carter Andrews, 2014).

A deep examination of issues of power in relation to race, class, and other forms of oppression in the education system is necessary to understand how to move towards equity in education and provide socially just pedagogy. Teachers, by engaging in critical self-reflection, can examine and challenge the assumptions behind their teaching practices – for instance, by examining their teaching practices from the perspective of a privileged social location, most often white, middle class. Developing racial consciousness or developing a "deeper awareness and understanding of race" enables teachers to reimagine knowledge transmission in teaching practices (Australian Institute of Teaching and School Leadership [AITSL], 2020, p. 19; see also Miller, 2013). Developing an awareness of educator identity and positioning within an education system is fundamental to seeing how positions of privilege can be used to transform pedagogic practices.

Moreover, since my Western perspective predisposes me to ignore "the ways in which colonialism and racism are reproduced" (Higgins et al., 2015, p. 252) in teaching practices, I may inadvertently continue the "subjugation of a student's identity" (Fogarty et al., 2015, p. 8). Western teachers may find it difficult to avoid the transmission of cultural norms in their teaching practices. However, at the site where Western and Yolŋu mathematics knowledges merge, different ways of thinking can flow like the currents mixing fresh and salt water (M. Ganambarr-Stubbs, personal communication, July 10, 2018). These currents bubble away underneath the surface. Similarly, developing a critical consciousness is like Ga<u>n</u>ma. The professional development work of teachers in an intercultural space includes a critical examination of new ideas at the site where different ways of thinking are exchanged. At the site, the exchange of ideas can occur in the first language and culture of the learners as well as the school-based language and culture focused on. At the site, students can move more freely, as if through a passage in a house leading to different rooms (Hogue, 2019).

4.2 *View of the Students*

Carter Andrews highlights how "meritocracy is a myth" (2014, 9:27). That is, meritocracy or "pulling up your bootstraps" works with the students who come to school with "the social and cultural codes" expected for students to learn at school (Carter Andrews, 2014, 9:44). The cultural norms of those who are privileged are favoured within the education system. However, other students who are unaware of these social and cultural codes when they start school are at a significant disadvantage. By Year 9, "Indigenous remote students are about five years behind non-Indigenous students" (Wilson, 2014, pp. 54–55). Students who do not embody or know these social and cultural codes (of privilege) are disadvantaged by an education system that perpetuates "differential access to forms of knowledge and power" (Smyth & Wrigley, 2013, p. 24).

In Australia, meritocracy is also redundant in remote Indigenous education. In remote Australian Indigenous contexts, as with most Organisation for

Economic Co-operation and Development (OECD) countries, school leaders and teachers are often inadequately prepared to navigate the unique needs of low-performing (segregated) disadvantaged schools (OECD, 2012). Across OECD countries, students are twice as likely to be low performing if they are from "disadvantaged socio-economic backgrounds" (OECD, 2012, p. 170). Remoteness confounded by "entrenched intergenerational socio-economic disadvantage" (Fogarty et al., 2015, p. 6) presents significant barriers to educational achievement for remote Indigenous students. Irrespective of race, nomadic family groups that are "difficult to reach" – for instance, Irish Travellers, European Roma, or New Zealand Māori – experience significant barriers to equitable educational achievement (OECD, 2012, p. 23).

Further, students who have not learned how to navigate the linguistic and cultural codes favoured in the Australian education system, and which are assessed through standardised testing by the National Assessment Program – Literacy and Numeracy (NAPLAN), are at a distinct disadvantage (Rigney et al., 2020). NAPLAN results are annually recorded for each Australian student in Year 3, 5, 7, and 9 as a numerical measure of student success in literacy (reading, writing, and language conventions – spelling, grammar, and punctuation) and numeracy taught according to the Australian Curriculum (ACARA, 2021). Relying on NAPLAN or other standardised assessments of teaching and learning effectiveness is problematic, particularly for Indigenous students "in remote areas" who are not familiar with the linguistic and cultural codes in these assessments (Wigglesworth et al., 2011, p. 340).

4.3 *View of the Curriculum*

Middle years (Years 8–9) students are at risk of disengaging from the mathematics curriculum if they are expected to solve a computational algorithm without a suitable context from which to consider the translation from the visual, oral, and numeric representations of a problem. The ability for students to shift easily from one representation to another is an essential foundation for the skills and understanding required when regrouping and renaming in computational algorithms (Payne & Rathmell, 1977, p. 152). For instance, by Year 8, Australian students are expected to flexibly regroup and rename base representations to "solve everyday problems involving rates, ratios and percentage" (ACARA, 2014, p. 17). The Australian Curriculum suggests example ways to assess student facility to flexibly regroup, rename, and translate between representation forms when computing algorithms (ACARA, 2014).

Figure 13 demonstrates how a middle years' student is solving an everyday problem, the percentage of akatyerr vegetation cover observed in the quadrat

FIGURE 13 Proportional reasoning – solving the percentage of plant cover in the quadrat frame
PHOTOGRAPH BY TIM WARE

frame being focused on, by flexibly processing oral, pictorial, written, and other representations of a proportion of an array. The student is sharing in Alyawarr and English an oral representation of the proportion of the array, while drawing and writing the percentage cover in a table and a graph. This lesson enhanced the mathematics meaning constructed in a remote Indigenous context since the lesson delivered embedded deliberate choices to enhance how language (Alyawarr and mathematics), visual, and numeric systems intersect (Harper & Parkin, 2019). Assessing and building numeracy skills, and creating learning from this understanding and in relation to students' lived experiences, promotes student engagement. Providing opportunities for students to experience learning flexibly is more equitable than providing curriculum that only focuses on procedural instruction and ignores a student's facility in numeracy.

4.4 *View of Knowledge*

Pedagogical decision-making that seeks to integrate students' real-life experiences while respecting the rights of Indigenous peoples has been given limited attention in the literature. Indigenous education research tends to question the ways in which Indigenous cultural knowledge concepts can be included in mathematics curriculum texts (see Dickenson-Jones, 2008) rather than the ways in which mathematics curriculum can be included in the Indigenous language and cultural context of the learner. At a site, concepts of measurement can be explored "in, with and as" an investigation occurs (Lloyd et al., 2016). For instance, assistant teacher, Sharmic, staff, students and I collected and sorted akatyerr at a site well known for its harvest capacity. Akatyerr harvesting practices occurred alongside Sharmic's guidance. Sharmic's handful of akatyerr collected was recorded, the time taken for the entire yield to be collected was recorded, and collections of dried fruit were weighed. The assistant teacher's knowledge of this cultural practice at this site was the basis for this investigation.

Culturally appropriate curriculum and pedagogic practices need to be developed in consultation with Indigenous staff and students. This collaboration demonstrates an awareness of the struggle of Indigenous people for self-determination (a solidarity) and challenges teachers to acknowledge how they may (unintentionally) be complicit in maintaining societal norms in teaching practices (Applebaum, 2010). Consultation with Indigenous community members is critical since it is difficult for Western teachers to integrate Indigenous perspectives into curriculum when a teacher's Western perspective means they "simultaneously engage in and resist" the "ways in which colonialism and racism are reproduced" (Higgins et al., 2015, p. 252) in Indigenous education. Additionally, when developing culturally responsive curriculum and pedagogy, it is important to consider where the points of convergence are between Western and Indigenous knowledge to facilitate border crossing (Aikenhead, 1997) between the two worldviews. The argument for a grounding of school curriculum in the local Indigenous context is gathering support (Battiste, 1998, 2013; Chinn, 2007; Meaney et al., 2013). Battiste (2013) has focused on the rights of First Nations peoples to enact treaties in relation to self-determination in education in general. Like the Indigenist principles of Rigney (1997, 1999) discussed in Section 2, Battiste (2013) provides a framework for inclusive context-based education; however, she does not provide guidelines for how it can be conducted in the classroom.

Indigenist research principles (Rigney, 1997, 1999) reflect the limitations of research traditions that have sought to privilege Indigenous voice in Indigenous education. For instance, the radical research tradition within Indigenous education is problematic since it unquestioningly accepts the value of Indigenous voice "for its power to transform" and connect the researcher to a place of "pre-colonial" origin (McConaghy, 2000, p. 245). The problem emerges with this romanticised and oversimplified notion that colonial power relations will be reversed by replacing non-Indigenous voices with Indigenous voices. Indigenous voice cannot be inserted into "curriculum, policy and practice"; rather "spaces for Indigenous pedagogies in Indigenous education" (McConaghy, 2000, p. 244) need to be made.

References

Aikenhead, G. (1997). Toward a first nations cross-cultural science and technology curriculum. *Science Education, 81*(2), 217–238. https://doi.org/10.1002/(SICI)1098-237X(199704)81:2%3C217::AID-SCE6%3E3.0.CO;2-I

Aikenhead, G. (2017). A 21st century economic, educational and ethical mathematics curriculum policy. *The Mathematics Enthusiast, 14*(1–3), 563–574. Retrieved January 11, 2023, from https://scholarworks.umt.edu/tme/vol14/iss1/29

Aikenhead, G., & Michell, H. (2011). *Bridging cultures: Indigenous and scientific ways of knowing nature*. Pearson Education Canada.

Applebaum, B. (2010). *Being white, being good: White complicity, white moral responsibility, and social justice pedagogy*. Lexington Books.

Arbon, V. (2008). *Arlathirnda ngurkarnda ityirnda: Being-knowing-doing: Decolonising Indigenous tertiary education*. Post Pressed.

Australian Curriculum, Assessment and Reporting Authority. (2014). *Australian curriculum: Work sample portfolio – Mathematics Year 8: Above satisfactory*. Retrieved January 11, 2023, from https://docs.acara.edu.au/curriculum/worksamples/Year_8_Mathematics_Portfolio_Above.pdf

Australian Curriculum, Assessment and Reporting Authority. (2020). *Australian curriculum: Mathematics (Version 8.4) – Year 8 level description*. Retrieved January 11, 2023, from https://bit.ly/3GB8zM7

Australian Curriculum, Assessment and Reporting Authority. (2021). *NAPLAN –general: What is NAPLAN?* Retrieved January 11, 2023, from https://www.nap.edu.au/naplan/faqs/naplan--general#:~:text=The%20National%20Assessment%20Program%20%E2%80%93%20Literacy,that%20all%20Australian%20children%20undertake

Australian Government. (2021). *Native title act 1993*. Office of Parliamentary Counsel Federal Register of Legislation. Retrieved January 11, 2023, from https://www.legislation.gov.au/Details/C2021C00450

Australian Government Department of Prime Minister and Cabinet. (2020). *Closing the gap report 2020*. Commonwealth of Australia. Retrieved January 11, 2023, from https://ctgreport.niaa.gov.au/sites/default/files/pdf/closing-the-gap-report-2020.pdf

Australian Institute of Teaching and School Leadership. (2020). *Indigenous cultural competency in the Australian teaching workforce: Discussion paper*. Retrieved January 11, 2023, from https://www.aitsl.edu.au/docs/default-source/comms/cultural-competency/aitsl_indigenous-cultural-competency_discussion-paper_2020.pdf

Baggett, H. C., Dunn, A. H., & Sondel, B. L. (2020). Critical teacher responsibility in tumultuous times: Engaging in a community of practice. In L. Parson & C. Ozaki (Eds.), *Teaching and learning for social justice and equity in higher education* (pp. 157–178). Palgrave Macmillan. https://doi.org/10.1007/978-3-030-44939-1

Battiste, M. (1998). Enabling the autumn seed: Toward a decolonized approach to Aboriginal knowledge, language, and education. *Canadian Journal of Native Education, 22*(1), 16–27.

Battiste, M. (2005). Indigenous knowledge: Foundations for first nations. *WINHEC: International Journal of Indigenous Education Scholarship, 1*, 1–17. Retrieved January 11, 2023, from https://journals.uvic.ca/index.php/winhec/article/view/19251

Battiste, M. (2013). *Decolonizing education: Nourishing the learning spirit.* Purich Publishing Limited.

Bishop, R., Berryman, M., Cavanagh, T., & Teddy, L. (2007). *Te Kotahitanga phase III whanaungatanga: Establishing a culturally responsive pedagogy of relations in mainstream secondary school classrooms.* New Zealand Ministry of Education. Retrieved January 7, 2023, from https://www.educationcounts.govt.nz/publications/schooling2/workforce/9977/9454

Brereton, M., Roe, P., & Schroeter, R. (2014, April 26–May 1). Beyond ethnography: Engagement and reciprocity as foundations for design research out here. In *CHI'14: Proceedings of the SIGCHI conference on human factors in computing systems, Toronto, ON, Canada* (pp. 1183–1186). https://doi.org/10.1145/2556288.2557374

Buckley, J. B., & Quaye, S. J. (2016). A vision of social justice in intergroup dialogue. *Race Ethnicity and Education, 19*(5), 1117–1139. https://doi.org/10.1080/13613324.2014.969221

Carter Andrews, D. J. (2014, March 11). *The consciousness gap in education – an equity imperative* [Video]. TEDx Talks. Retrieved January 11, 2023, from https://www.youtube.com/watch?v=iOrgf3wTUbo

Carter Andrews, D. J., Bartell, T., & Richmond, G. (2016). Teaching in dehumanizing times: The professionalization imperative. *Journal of Teacher Education, 67*(3), 170–172. https://doi.org/10.1177/0022487116640480

Central Land Council. (2017). *Central Land Council annual report 2016–17.* Retrieved December 9, 2022, from https://www.clc.org.au/wp-content/uploads/2021/03/CLC-Annual-Report-2016-2017.pdf

Chinn, P. (2007). Decolonizing methodologies and Indigenous knowledge: The role of culture, place and personal experience in professional development. *Journal of Research in Science Teaching, 44*(9), 1247–1268. https://doi.org/10.1002/tea.20192

Christie, M., Guyula, Y., Gotha, K., & Gurruwiwi, D. (2010). The ethics of teaching from country. *Australian Aboriginal Studies, 2*, 69–80. Retrieved January 11, 2023, from https://search.informit.org/doi/abs/10.3316/ielapa.767045875963420

Davis, J., & Grose, S. (2008). *Which way? What happens when embedding Aboriginal and Torres Strait Islander perspectives in schools meets the professional standards for teachers and an accountability matrix?* Stronger Smarter Institute. Retrieved January 11, 2023, from https://strongersmarter.com.au/wp-content/uploads/2020/08/Which_Way__by_John_Davis_and_Sharon_Grose_for_Wipce_2008.pdf

Dickenson-Jones, A. (2008). Transforming ethnomathematical ideas in Western mathematics curriculum texts. *Mathematics Education Research Journal, 20*(3), 32–53. https://doi.org/10.1007/BF03217529

Dobson, P. V., Walsh, F., & Sati, W. (2008). *Anpernirrentye: Relationships between bush foods, creation laws, people, country and all things, illustrated by three plants* [DVD]. Desert Knowledge Cooperative Research Centre.

Douglas, J. (2011). *Learning from country: The value of country visits in remote schools for community engagement and natural and cultural resource management* (Desert Knowledge Cooperative Research Centre Report No. 69). Ninti One Limited. Retrieved January 11, 2023, from https://www.nintione.com.au/resource/NintiOneResearchReport_69_Thevalueofcountryvisitsinremoteschools.pdf

Education Council. (2019). *Alice Springs (Mparntwe) education declaration.* Retrieved January 11, 2023, from https://nla.gov.au/nla.obj-2391375250/view

Ewert, G. D. (1991). Habermas and education: A comprehensive overview of the influence of Habermas in educational literature. *Review of Educational Research, 61*(3), 345–378. https://doi.org/10.3102/00346543061003345

Ewing, B. (2011). Direct instruction in mathematics: Issues for schools with high Indigenous enrolments: A literature review. *Australian Journal of Teacher Education, 36*(5), 65–92. https://doi.org/10.14221/ajte.2011v36n5.5

Fogarty, W., Lovell, M., & Dodson, M. (2015). A view beyond review: Challenging assumptions in Indigenous education development. *UNESCO Observatory Multi-Disciplinary Journal in the Arts, 4*(2), 1–23. Retrieved October 14, 2022, from https://hdl.handle.net/1885/13420

Foley, D. (2003). Indigenous epistemology and Indigenous standpoint theory. *Social Alternatives, 22*(1), 44–52. Retrieved January 11, 2023, from https://www.academia.edu/907693/Indigenous_Epistemology_and_Indigenous_Standpoint_Theory

Gough, M. (2008). The changing relationship between first nations people and museums. In C. Avery & D. Fichter (Eds.), *kā-kī-pē-isi-nakatamākawiyahk* [Our legacy: Essays] (pp. 205–232). University of Saskatchewan. Retrieved January 11, 2023, from http://scaa.sk.ca/ourlegacy/essays/OurLegacy_Essays.pdf

Green, J., Blackman, D., & Moore, D. (2019). *Alyawarr to English dictionary* (2nd ed.). IAD Press.

Guenther, J., Osborne, S., Arnott, A., & McRae-Williams, E. (2017). Hearing the voice of remote Aboriginal and Torres Strait Islander training stakeholders using research methodologies and theoretical frames of reference. *Race Ethnicity and Education, 20*(2), 197–208. https://doi.org/10.1080/13613324.2015.1110294

Halbert, J., & Kaser, L. (2019). A focus on equity and excellence in networks of inquiry. *The Queensland Principal, September*, 21–23. Retrieved January 11, 2023, from https://noiie.ca/wp-content/uploads/2019/10/Halbert-and-Kaser-The-Qld-Principal-Sep2019.pdf

Harper, H., & Parkin, B. (2019). Talking and writing to develop mathematical meanings in a remote Indigenous context. In J. Rennie & H. Harper (Eds.), *Literacy education and Indigenous Australians* (pp. 109–125). https://doi.org/10.1007/978-981-13-8629-9_7

Harris, P. (1980). *Measurement in tribal Aboriginal communities*. Northern Territory Department of Education.

Higgins, M., Madden, B., & Korteweg, L. (2015). Witnessing (halted) deconstruction: White teachers' "perfect stranger" position within urban Indigenous education. *Race, Ethnicity and Education, 18*(2), 251–276. https://doi.org/10.1080/13613324.2012.759932

Hogue, M. (2019). Two-eyed seeing for both ways knowing: Using digital technology to create liminal spaces for Indigenous STEM success. *Discussions on University Science Teaching: Proceedings of the Western Conference on Science Education, 2*(1). Retrieved January 11, 2023, from https://ojs.lib.uwo.ca/index.php/wcsedust/article/view/8110

House of Representatives Standing Committee on Aboriginal and Torres Strait Islander Affairs. (2012). *Our land our languages: Language learning in Indigenous communities*. Parliament of the Commonwealth of Australia. Retrieved January 29, 2023, from https://www.aph.gov.au/parliamentary_business/committees/house_of_representatives_committees?url=atsia/languages2/report.htm

Innes, R. A. (2009). "Wait a second. Who are you anyways?" The insider/outsider debate and American Indian studies. *American Indian Quarterly, 33*(4), 440–461.

Janke, T. (2019, April 10). *ICIP True Tracks® cultural protocols plan* [Workshop]. Terri Janke and Company.

Janke, T., & Cubillo, C. (2019). *Towards an ICIP (Indigenous cultural and intellectual property) strategy for your work at Walyarra School* [pseudonym]. Terri Janke and Company.

Kendall, E., Sunderland, N., Barnett, L., Nalder, G., & Matthews, C. (2011). Beyond the rhetoric of participatory research in Indigenous communities: Advances in Australia over the last decade. *Qualitative Health Research, 21*(12), 1719–1728. https://doi.org/10.1177/1049732311418124

Ladson-Billings, G. (1995). Toward a theory of culturally relevant pedagogy. *American Educational Research Journal, 32*(3), 465–491. https://doi.org/10.3102/00028312032003465

Langton, M. (2013). Indigenous exceptionalism and the constitutional "race power". In *Future leaders, space place & culture*. Retrieved January 11, 2023, from https://www.futureleaders.com.au/book_chapters/pdf/Space-Place-Culture/Marcia-Langton.pdf

Lloyd, K., Suchet-Pearson, S., Wright, S., Burarrwanga, L., Ganambarr, R., Gunambarr-Stubbs, M., Ganambarr, B., & Maymuru, D. (2016). Morrku Mangawu—Knowledge on the Land: Mobilising Yolŋu mathematics from Bawaka, North East Arnhem Land, to reveal the situatedness of all knowledges. *Humanities*, *5*(3), 1–13. https://doi.org/10.3390/h5030061

Lyotard, J. F. (1986). *The postmodern condition: A report on knowledge* (G. Bennington & B. Massumi, Trans.). University of Minnesota Press. (Original work published 1984)

Lyotard, J. F. (1988). *The differend: Phrases in dispute* (G. Van Den Abbeele, Trans.). Manchester University Press. (Original work published 1983)

Madusise, S. (2015). Cultural villages as contexts for mediating culture and mathematics education in the South African curriculum. *Revista Latinoamericana de Etnomatemática*, *8*(2), 11–31. Retrieved January 11, 2023, from https://revista.etnomatematica.org/index.php/RevLatEm/article/view/203

Martin, K. (2008). *Please knock before you enter: Aboriginal regulation of outsiders and the implications for researchers*. Post Pressed.

Maynard, J. (2007). Circles in the sand: An Indigenous framework of historical practice. *The Australian Journal of Indigenous Education*, *36*(S1), 117–120. https://doi.org/10.1017/S1326011100004786

McConaghy, C. (2000). *Rethinking Indigenous education: Culturalism, colonialism, and the politics of knowing*. Post Pressed.

Meaney, T., Trinick, T., & Fairhall, U. (2013). One size does NOT fit all: Achieving equity in Māori mathematics classrooms. *Journal for Research in Mathematics Education*, *44*(1), 235–263. https://doi.org/10.5951/jresematheduc.44.1.0235

Merne Altyerre-ipenhe [Food from the Creation time] Reference Group, Douglas, J., & Walsh, F. (2011). *Aboriginal people, bush foods knowledge and products from central Australia: Ethical guidelines for commercial bush food research, industry and enterprises* (Desert Knowledge Cooperative Research Centre Report No. 71). Ninti One Limited. Retrieved January 11, 2023, from https://www.nintione.com.au/resource/NintiOneResearchReport_71_BushFoodGuidelines.pdf

Miller, M. (2013). *Action for change? Embedding Aboriginal and Torres Strait Islander perspectives in early childhood education curricula* [PhD thesis, Queensland University of Technology]. QUT ePrints. Retrieved January 11, 2023, from https://eprints.qut.edu.au/60905/

Mills, K. A., Sunderland, N., & Davis-Warra, J. (2013). Yarning circles in the literacy classroom. *The Reading Teacher*, *67*(4), 285–289. https://doi.org/10.1002/trtr.1195

Morrison, A., Rigney, L., Hattam, R., & Diplock, A. (2019). *Toward an Australian culturally responsive pedagogy: A narrative review of the literature*. University of South Australia. Retrieved December 9, 2022, from https://apo.org.au/node/262951

Munns, G., & Woodward, H. (2006). Student engagement and student self-assessment: The REAL framework, assessment in education. *Principles, Policy & Practice, 13*(2), 193–213. https://doi.org/10.1080/09695940600703969

National Native Title Tribunal. (2014). *Native title determination details: DCD2014/010 – Grandplover* [pseudonym]. National Native Title Tribunal.

Northern Territory Board of Studies. (2017). *Keeping Indigenous languages and culture strong: A plan for the teaching and learning of Indigenous languages and cultures in Northern Territory Schools*. Northern Territory Government.

Northern Territory Department of Education. (2017). *NT Indigenous languages and cultures: First Languages Maintenance (L1M)*. Northern Territory Government. Retrieved January 11, 2023, from https://education.nt.gov.au/__data/assets/pdf_file/0020/715232/ntilc-l1m-nov18-03.pdf

Northern Territory Department of Education. (2018). *Indigenous education strategy – Issue 19*. Retrieved January 11, 2023, from https://education.nt.gov.au/statistics-research-and-strategies/indigenous-education-strategy/indigenous-education-strategy-issue-19/communities-empowered-in-school-decisions

Northern Territory Department of Education. (2021). *Indigenous education strategy employment pathways: Communication and problem solving: Numeracy – Area (version 2.2)*. Northern Territory Government. Retrieved January 11, 2023, from http://ed.ntschools.net/partpath/team/epprograms/SiteAssets/SitePages/Communication%20and%20Problem%20Solving/Areav2.2.pdf

Organisation for Economic Co-operation and Development. (2012). *Equity and quality in education: Supporting disadvantaged students and schools*. OECD Publishing. https://doi.org/10.1787/9789264130852-en

Payne, J., & Rathmell, E. (1977). Numeration. In J. Payne (Ed.), *Mathematics learning in early childhood: 1977 yearbook of the National Council of Teachers of Mathematics*. NCTM.

Perso, T., & Hayward, C. (2015). *Teaching Indigenous students: Cultural awareness and classroom strategies for improving learning outcomes*. Allen & Unwin.

Peters, M. (1995). Education and the postmodern condition: Revisiting Jean-François Lyotard. *Journal of Philosophy of Education, 29*(3), 387–400. https://doi.org/10.1111/j.1467-9752.1995.tb00367.x

Rigney, L. I. (1997). *Racism and physical education: A critical Indigenist analysis of the Senate Standing Committee's report on physical and sport education* [Unpublished master's thesis]. University of South Australia.

Rigney, L. I. (1999). Internationalization of an Indigenous anticolonial cultural critique of research methodologies: A guide to Indigenist research methodology and its principles. *Wicazo Sa Review, 14*(2), 109–121. https://doi.org/10.2307/1409555

Rigney, L. I. (2017, September 29). Decolonising Pacific schools: Toward an Australian culturally-responsive pedagogy? [Video]. YouTube. Retrieved December 8, 2022, from https://www.youtube.com/watch?v=bpqcU27kkWc

Rigney, L. I. (2019a, September 20). A time of change and "super diversity" in early childhood education. *The Spoke: Early Childhood Australia's Blog*. Retrieved December 19, 2022, from https://thespoke.earlychildhoodaustralia.org.au/time-change-super-diversity-early-childhood-education/

Rigney, L. I. (2019b, September 25–28). *Early years practice using Aboriginal learner cultural strengths? Culturally responsive pedagogies in dialogue with Reggio Emilia* [Australian keynote address PowerPoint slides]. Early Childhood Australia National Conference 2019, Hobart, Tasmania, Australia. Retrieved December 19, 2022, from https://www.ecaconference.com.au/2019/presenter-papers-2019/

Rigney, L. I., Garrett, R., Curry, M., & MacGill, M. (2020). Culturally responsive pedagogy and mathematics through creative and body-based learning: Urban Aboriginal schooling. *Education and Urban Society*, *52*(8), 1159–1180. https://doi.org/10.1177/0013124519896861

Rigney, L. I., & Hattam, R. (2020, July 2). *Professor Lester-Irabinna Rigney & Professor Robert Hattam: Toward an Australian culturally responsive pedagogy* [Video]. YouTube. Retrieved October 12, 2022, from https://www.youtube.com/watch?v=PxwNDTDEWes

Scherer, P., & Steinbring, H. (2006) Noticing children's learning processes – Teachers jointly reflect on their own classroom interaction for improving mathematics. *Journal of Mathematics Teacher Education*, 9, 157–185. https://doi.org/10.1007/s10857-006-0004-7

Scheurich, J., & Young, M. (1997). Coloring epistemologies: Are our research epistemologies racially biased? *Educational Researcher*, *26*(4), 4–16. https://doi.org/10.2307/1176879

Siemon, D., Breed, M., & Virgona, J. (2005). From additive to multiplicative thinking: The big challenge of the middle years. In J. Mousley, L Bragg, & C. Campbell (Eds.), *Mathematics – celebrating achievement: Proceedings of the 42nd conference of the Mathematical Association of Victoria*. MAV.

Smith, L. T. (1999). *Decolonizing methodologies: Research and Indigenous peoples*. Zed Books.

Smyth, W., & Wrigley, T. (2013). *Living on the edge: Rethinking poverty, class and schooling*. Peter Lang.

Turner, M. (2005). *Bush foods: Arrernte foods from Central Australia* (2nd ed.). IAD Press.

Waldby, C. (1995). Feminism and method. In B. Caine & R. Pringle (Eds.), *Transitions: New Australian feminisms* (pp. 15–28). Allen & Unwin.

Wigglesworth, G., Simpson, J., & Loakes, D. (2011). NAPLAN language assessments for Indigenous children in remote communities: Issues and problems. *Australian Review of Applied Linguistics*, *34*(3), 320–343. https://doi.org/10.1075/aral.34.3.04wig

Wilson, B. (2014). *A share in the future. Review of Indigenous education in the Northern Territory.* The Education Business. Retrieved January 11, 2023, from https://www.nt.gov.au/__data/assets/pdf_file/0020/229016/A-Share-in-the-Future-The-Review-of-Indigenous-Education-in-the-Northern-Territory.pdf

Woodward, A. (n.d.). Jean-François Lyotard (1924–1998). In *Internet encyclopedia of philosophy*. Retrieved June 20, 2021, from https://iep.utm.edu/lyotard/#H11

Woolley, T., Sivamalai, S., Ross, S., Duffy, G., & Miller, A. (2013). Indigenous perspectives on the desired attributes of medical graduates practising in remote communities: A northwest Queensland pilot study. *The Australian Journal of Rural Health, 21*(2), 90–96. https://doi.org/10.1111/ajr.12018

CHAPTER 3

Embedding Living Stories in Mathematics Education

On 31 May 2017 I attended a bush trip with Alyawarr and Anmatyerr woman harvesters and students. Accompanying us was Ngangiwumirr/Arrernte bush food caterer Rayleen Brown (Janke, 2021, p. 218). This was the first time I was shown a handful of akatyerr. As she held out the desert fruit in her hand, Rayleen shared with the students:

> How important it is that young people learn about these special plants from the old ladies … This plant has got plenty of iron to make your blood strong, vitamin D and plenty of vitamin C … If you can eat a little bit of bush tucker [akatyerr, wattle seed] everyday it is going to be good for your stomach, good for your skin, and good for your blood, make you healthy. So *really important* that you learn from the old people and listen. (R. Brown, personal communication, May 31, 2017)

Rayleen's words echo what many others have said about the importance of young people learning traditional knowledge and skills from the elders during wild-harvest activities. This is especially important to continue to preserve cultural protocols and customary law related to harvest practices (Merne Altyerre-ipenhe Reference Group et al., 2011; Walsh & Douglas, 2011). The "long term sustainability" of these special plants in the bush food industry depends on ongoing intergenerational knowledge sharing on bush trips such as these (Maclean et al., 2019, p. 53).

1 Utopia Community

After substantial rain in the cool time of 2017, piles of sandy red soils bordered either side of the graded roads near where I lived in the Central Desert region. As I walked along these familiar roads, I began noticing the presence of akatyerr growing in the disturbed soils. I took photos such as the one pictured in Figure 14, which shows both unripe green fruit and ripe yellow fruit.

After noticing akatyerr growing in graded disturbed soils and having listened to wild-harvest industry advocate Rayleen Brown speak of the health benefits as

 | DOI:10.1163/9789004682757_003

FIGURE 14 Akatyerr I noticed growing in disturbed roadside sandy red soils near where I lived. The yellow fruit are ripe, the green fruit are unripe
PHOTOGRAPH BY NICOLE BOYD

well as the economic, spiritual, and cultural values contained within a handful of this nutritious fruit, I began conceptualising my education research project based on this product. Soon after, I submitted my first abstract for a higher degree research conference. At the same time, as a remote secondary teacher, I was grappling with the "watered down curriculum" mandated in the numeracy program (Wilson, 2014, p. 136). When students reach Year 8, their families can choose if they want their children to stay in the community or homeland where they live to continue their schooling at the local school. If this option is chosen, students can attend classes for "post primary literacy and numeracy and [if offered] an employment pathways program … which will provide skills … to gain employment" (NTG, 2017, para. 4). A model for delivering "remote employment pathways" was recommended for very remote Indigenous schools such as ours, where "student numbers [were] not sufficient to deliver a full secondary education curriculum" (NTDOE, 2016b, "Secondary Pathways" section, para. 4). This model enabled local community to inform the local content for delivering "clear vocational pathways to local employment" (NTDOE, 2016b, "Secondary Pathways" section, para. 4). Our school held meetings between teaching staff, the principal, Indigenous school leaders, and assistant teachers before students began a locally contextualised "Big Business" unit of work (NTDOE, 2016a). The purpose of these meetings was to seek advice from the community and school council to identify what local content needed to be included to make learning experiences relevant to students (NTDOE, 2016c). By August 2017, I presented our school's Employment Pathways proposed unit of work at a regional workshop (Tennant Creek) within the NTDOE. The local content focused on was "exploring opportunities to support the next generation of harvesters to sustainably develop the desert raisin wild harvest industry using school curriculum initiatives" (Boyd, 2017).

1.1 *First Language Use*

My awareness of the importance of recentring on first languages in the Employment Pathways curriculum planning emerged through the study. I was not aware of the importance of first languages to support student identity when I first developed the Big Business unit plan. The workshop had provided an opportunity for me to present initial ideas, some of which were distilled from the Utopia community plans "to improve the lives of children" as established in the Apmer Akely-akely committee and discussed at school council meetings (CAYLUS, 2017). School council meetings provided a forum to discuss the importance of first language and culture lessons. Ongoing discussions and engagement with concepts of nation (re)building (Cornell, 2006) and acts of resurgence (Corntassel, 2012) led to my understanding of the importance of recentring on the language through which the Australian mathematics curriculum was taught. Speaking first languages during a place-based mathematics problem-solving lesson is one way to practise "acts of resurgence" (Corntassel, 2012, p. 88). Through first-language use, speakers are able to perceive and respond to interconnections they have to the land, and the plants and animals that belong to that land, through their understanding of kinship rules (see Green et al., 2019, p. 422). Language speaking reaffirms the "relational, place-based existence" of First Nations peoples (Corntassel, 2012, p. 88). Everyday practices such as these renew connections to land, culture, and community (Corntassel, 2012, p. 89). Continuously renewing connections is necessary since "yonega (White settler) encroachment onto Indigenous homelands and waterways" has disrupted the "complex spiritual, political and social relationships" embedded in everyday practices for First Nations people in a Canadian context (Corntassel, 2012, p. 89). As Mohawk scholar Taiaiake Alfred (2009) writes, daily acts of renewed connection are necessary since colonialism has resulted in:

> a multigenerational and multifaceted process of forced dispossession and attempted acculturation – a disconnection from land, culture and community – that has resulted in political chaos and social discord within First Nations communities and the collective dependency of First Nations upon the state. (p. 52)

As an NTG teacher and agent of the colonial state, I could "personally and collectively" support daily acts of resurgence (Corntassel, 2012, p. 89) through creating opportunities for Indigenous languages to be spoken in everyday aspects of teaching mathematics. Indigenous language curriculum could be offered within NTG schools with the caveat that it "must be led by Indigenous elders and community members who are the custodians of the languages and cultures" (NTBOS, 2017, p. 4) and differentiated to suit the learner's pathway to language acquisition (NTBOS, 2017, p. 51). I facilitated the "creation of spaces and places where Aboriginal voices are heard" (Watson, 2018, p. 17) and reflected in the school curriculum development. As an ally, I relayed the community's aspiration to build opportunities for local employment based on cultural and local knowledge and local language (CAYLUS, 2017) so that the learner's Indigenous language needs could be incorporated in the curriculum-writing process. In this sense, we "de-centre[d] colonial power" to counteract education practices which focus on "assimilation into a dominant paradigm with 'progress' at the centre" (Watson, 2018, pp. 8–16). Rather than conform to the direction of "policymakers who frame new government initiatives as 'economic development'", we reframed the Employment Pathways initiative to explore the opportunity to make "larger connections" to "Indigenous economies linking homelands, cultures and communities" (Corntassel, 2012, p. 88). This involved creating a space where Aboriginal epistemologies could sit in the Australian mathematics curriculum and "re-centre[d] Aboriginal Peoples' laws and authority over place" in pedagogic practices (Watson, 2018, p. 8). This reflected a shared understanding that "responsibilities to culture, family, and the land – such as passing on traditional knowledge" (Wilson et al., 2019, p. 154) are prioritised over paid employment (see, for example, Altman, 2009). Mainstream education to employment pathways is understood as "inherently challenging" for Aboriginal and/or Torres Strait Islander peoples in remote communities, and "aspirations for belonging to family, community and country which shape economic engagement can be overlooked by a system that values different goal orientations and privileges certain ways of getting there" (McRae-Williams, 2014, p. 57).

The feedback I received from the presentation was to continue to develop a unit of work based on bush foods (and bush medicine) as the local content because it was a "perfect example of community and culture working together in and out of the classroom" (Kate Francis, personal communication, August 2017). Soon after, we began harvesting bush medicine plants and recording the plant knowledge in the unit of work. On one bush trip we harvested six large bags of ilpengk (*Eremophila dalyana*; Green et al., 2019, p. 215). This shrub, commonly known as gidgee fuchsia bush, grows to about 1.5 metres (Green et al., 2019, p. 215). The narrow grey-green leaves are harvested by hand or with secateurs. The branches were collected in garbage bags and stored at school.

Soon after, we crushed the leaves by hand (see Figure 15 in Section 1.3) and boiled the leaves in a billy with animal fat on a fire. When the mixture cooled, it was sieved and stored in jars to sell as a rubbing medicine.

1.2 *Australian Native Food and Botanicals* (*ANFAB*)

To further develop the students' Big Business and numeracy skills, on 28 February 2018, Indigenous language and culture school staff, assistant teachers, and I attended an Australian Native Food and Botanicals (ANFAB) forum. Our attendance at this event was supported by the school leaders and Apmer Akely-akely committee and Utopia bush food harvesters. We were also encouraged and supported to participate in the forum by a bush foods industry development officer, Donna Digby, from the Northern Territory Department of Primary Industries and Resources (Alice Springs). The focus of the forum was on bush foods and bush medicines harvested in Central Australia. Here, a panel of women harvesters were consulted and many delegates at the forum shared their wild harvest stories. We were prompted to a broader discussion around intellectual property (IP) relating to bush food products. Stories shared demonstrated the "inadequacy of IP laws to protect the interests of Indigenous peoples" when traditional knowledge is utilised in business products (Maclean et al., 2019, p. 55; see also Lingard & Martin, 2016; Logue et al., 2018). There are opportunities, however, to recognise traditional knowledge related to *place* – using geographic indicators, for instance, since "currently there is generally very limited acknowledgement of the contribution of Indigenous people, and geographic sources of bush foods, on product labelling" (Maclean et al., 2019, p. 56; see also Merne Altyerre-ipenhe Reference Group et al., 2011). In discussions, we understood there were many opportunities to "grow the grower" through an exploration of what it was like to harvest on country (ANFAB, n.d.). From listening to the stories shared, I realised there was a lot of scope to learn to grow, value-add and market native produce (ANFAB, n.d.) in our school context.

At this forum, I met ethno-ecologist Fiona Walsh. Fiona had co-authored *Angka Akatyerr- Akert: A Desert Raisin Report* with Josie Douglas and Alyawarr speakers from Ampilatwatja (2009). This text provides ideas for teachers to link science, Alyawarr and other languages, mathematics, seasons, and their interrelationships with akatyerr (e.g., pp. 48–49). Combined, the authors properly attributed the cultural knowledge interrelated to akatyerr to the "relevant Indigenous language group" and community (Janke, 2021, p. 25). They "named" and "acknowledged" Alyawarr speakers from Ampilatwatja as the "custodians of Indigenous cultural knowledge" contained within the report (Janke, 2021, p. 25). As "certain traditional owners" are known to "have special rights and responsibilities to this species", the rights to information shared by the custodians of the plants were protected by copyright notices (Alyawarr speakers from Ampilatwatja et al., 2009, p. iii).

1.3 *Indigenous Cultural Knowledge Acknowledgement*

The ANFAB forum led me to pursue more of Terri Janke's work. I was informed by Terri Janke's True Tracks TED talk (2016). I became aware of the risks of Indigenous cultural knowledge being exclusively controlled by "an inventor or organisation" who holds an approved patent under the *Patents Act 1990* (Cth) (Janke, 2021, p. 207). That is, a patent application could be approved based on evidence that an "inventive step" or "new method" of extracting a benefit from a plant can be demonstrated (Janke, 2021, p. 207). I sought to avoid becoming entangled in what Indigenous scholar Irene Watson refers to as a "colonial knot" (2018, p. 16). Irene's critical First Nations standpoint on "the law and politics of control and power" stems from her ancestors of the "Tanganekald, Meintangk and Boandik Peoples of the South-East of South Australia" (Watson, 2018, p. 7). She asserts that, since "our lands and our lives" are still under the control and power of the settler colonial state, "First Peoples are asked to agree … in a context in which they have no authority to say 'yes' and no power to say 'no'" (Watson, 2018, pp. 15–16). Mindful of the *colonial knot*, I began agreement-making discussions about proper attribution of cultural knowledge of bush medicine plants with my students and assistant teachers. I asked the students how they wanted to acknowledge the cultural knowledge contained within the bush balm products we developed through our "Pop-up Shop" within the Employment Pathways curriculum (NTDOE, 2016a). Students decided to name and acknowledge the place from where the bush medicine plants were collected after viewing Terri Janke's TED talk (2016) and discussing the "opportunity to use a geographical indication" when marketing the bush medicine products to "show the particular place the product comes from" (Maclean et al., 2019, p. 56).

By May 2018, students, assistant teachers, and I launched our first Pop-up Shop (NTDOE, 2016a) bush balm business in the school, at the local shop, and at school council meetings. We advertised our products and sold arreth (*Eremophila freelingi*; Green et al., 2019, p. 139), apmikw (*Callitris glaucophylla*; Green et al., 2019, p. 111), and ilpengk (*Eremophila dalyana*; Green et al., 2019, p. 215) to community members. This curriculum developed in consultation with Apmer Akely-akely complemented and was informed by the work that already existed in the community. Figure 15 shows students preparing the bush medicine plants.

After a bush medicine plant had been crushed (Figure 15), boiled, sieved, and cooled, it was later put into different-sized containers. This curriculum

FIGURE 15 Students prepare bush medicine plants for a Pop-up Shop as part of the Employment Pathways curriculum
SOURCE: NTDOE, 2016A; PHOTOGRAPH BY CARI BERTRAM

was advocated by the chair of the Apmer Akely-akely committee, Esau Nelson. Apmer Akely-akely, as finalists in the Reconciliation Australia (2019) National Indigenous Governance Awards, demonstrated how students operated our bush balm business. Students had learned from elders how to harvest and process cultural materials for the benefit of staying connected to their culture while encouraging self-employment.

1.4 *Bush Medicine*

This bush balm business provided an opportunity to begin the discussion on what was an appropriate price for three different-sized containers. Students developed numeracy skills in "number sense and algebra" to establish a price by comparing units (ratios, rates, and proportion) (ACARA, 2020e). Applying proportional reasoning to bush medicine products, students interpreted rates "as a relationship between two different types of quantities" (ACARA, 2020e) – the quantity of money ($) and the quantity of bush medicine (millilitres). When assigning prices that students thought were reasonable to sell to the community (see Figure 16), we discussed how "increasing one quantity in a ratio will affect the total proportion" or value of the product (ACARA, 2020e).

Students priced the bush balm products of different volumes according to what they calculated were fair prices: $8 for 80 mL, $15 for 175 mL, and $25

FIGURE 16 The teacher (me) eliciting a fair price from Year 8/9 students while displaying the first batch of bush medicine
PHOTOGRAPH BY PAUL EVANS

for 375 mL. Students also used their mathematical knowledge from the curriculum strand of Number and Algebra to "carry out the four operations with rational numbers and integers, using efficient mental and written strategies" (ACARA, 2020d, "Number and place value") when exchanging money in business transactions and recording sales and the stock available.

Income generated from the enterprise was modest; however, the skills of a small start-up business have been applied to many learning opportunities that exist on country in the community and at school. Students began to understand the value of the work they had undertaken. Although crushing bush medicine leaves was arduous (see Figure 15), students made $1,000 in this venture and assisted to support the health of community members who could not make the medicine for themselves.

1.5 *Original Artworks*

Following the success of the bush balm business, I was asked by the school leaders to design another Pop-up Shop unit of work to price and sell Utopia Desert Art. This enabled students to apply the same numeracy skills in number sense and algebra (applying proportional reasoning) to a new context. Students began to assess the value of local original artworks and price them in a catalogue. Utopia Desert Art products were priced by measuring the

Figure 17
Sharmic (centre right) and Henry (centre left) consult the catalogue created for the Utopia Desert Art Pop-up Shop
PHOTOGRAPH BY PAUL EVANS

area of each canvas and applying a rate ($/cm^2) for each artist showcased. By revealing the rate per square centimetre for famous Utopia artist Emily Kame Kngwarreye (deceased), students could understand the value of her art and the premium quality. Students could compare this rate to artworks created by famous contemporary Utopia artists and emerging artists. Sharmic and Henry, as assistant teachers (pictured in Figure 17), had overseen the Utopia Desert Art curriculum development.

After the rate was provided for each Utopia artist's work showcased, students then measured the area of each artwork and applied the price per square centimetre (rate) to each canvas. The students' calculations were used to price and advertise each artwork for a fair price. Figure 17 depicts cultural advisors and assistant teachers Sharmic and Graham consulting the catalogue produced as part of the school-based curriculum. School staff and I supported students to participate in collecting $10,000 in art sales of Utopia Desert Art over two days during the "Desart" festival held on 7–8 September 2018 (Desart, 2018). Students used numeracy skills to calculate change and handled cash transactions developed through micro-business experiences during the bush balm Pop-up Shops. Students learned how to communicate the story of the artists whose work was showcased and how to handle credit card and cash transactions with interstate and local visitors purchasing artworks, cultural artefacts, and packaged dried akatyerr. Figure 18 depicts a Year 9 student selling a bush medicine original artwork and completing the cash transaction. She is standing alongside the customer (image cropped), a local from Alice Springs, and holding the money collected in the transaction after checking the sale price. On the

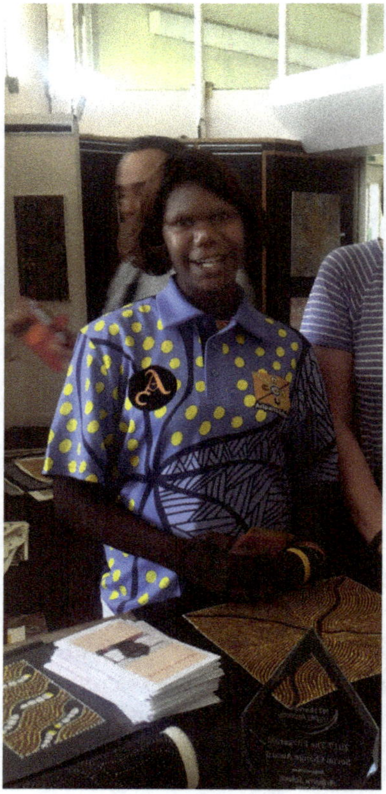

FIGURE 18
A Year 9 student is selling a bush medicine original artwork and completing the cash transaction while the customer (image cropped) stands alongside her
PHOTOGRAPH BY PAUL EVANS

table in front of the student is a brochure produced by her classmates which highlights the story of Utopia artists' artwork. Also on the table is the school's Human Rights Award.

1.6 *Significant Cultural and Ecological Sites*

There are many opportunities to apply proportional reasoning skills to address complex real-world problems at significant cultural and ecological sites in Utopia. During a preliminary study with limited sample size by Walsh and Douglas (2011), the akatyerr harvest rates (kilograms of dried fruit picked and sorted per hour) were measured, with results indicating an average harvest rate of 0.6 kg/h (n = 5 records) for a region in Utopia (p. 407). Additionally, plants considered invasive or vulnerable have also been monitored at culturally and ecologically significant sites in Utopia. In 2016, ranger work to construct a fence commenced to protect the remnant rainforest plant *Typhonium*, known locally as "pretty flower" (Louisa Long, personal communication, June 13, 2019; see Figure 19) from trampling by wandering animals such as horses

FIGURE 19 A *Typhonium* plant growing at the base of a river red gum tree (*Eucalyptus camaldulensis*) in an area downstream of the permanent water source. The broad leaves on this plant are unique for a desert ecosystem
PHOTOGRAPH BY TIM WARE

(Northern Territory Department of Environment, Parks and Water Security, 2021).

Inside the fenced area, the *Typhonium* is known to occur on the banks of the waterhole, as well as areas upstream or downstream of this permanent water source (Northern Territory Department of Environment, Parks and Water Security, 2021). The ranger group adjacent and with connections to this land (Anmatyerr), and traditional owners related by kinship to this land, had known of the environmental threats to this site, such as encroaching and invasive Buffel grass (*Cenchrus ciliaris*). The position I held in the community as a teacher provided a sustained relationship with traditional owners, who as local land managers would later (during 2022) form an emerging Utopia Alyawarr ranger group. Sustained relationships with traditional owners, harvesters, and students guided the research process. This was particularly important when co-designing curriculum opportunities to investigate the percentage vegetation cover of plants such as akatyerr ($\%/m^2$) and/or the students' harvest rate (kg/h) at the chosen culturally and ecologically significant investigation site. This investigation was critical to monitor the effect of site protection efforts to promote the growth and production of significant plants.

2 My Positionality – An Insider/Outsider

I was positioned as both an insider and an outsider in the Utopia community. I was shown this significant site by the students and community members to make observations and apply mathematical skills to further our understanding of how to protect the ecological and cultural significance held in stories and observations of this country. I examined, reflexively, how my story as the mathematics teacher, and my story as the researcher, highlighted ways to remove barriers for female Indigenous students' progression in measurement to address the following research question:

> How does a non-Indigenous teacher contextualise female Indigenous students' interests and lived experiences of harvesting akatyerr when teaching measurement?

The research focus enabled me to view students' lived experiences through a strengths discourse and liberating perception of potentiality. This focus created opportunities for students to engage in substantive conversation in their first language and use this to translate to the school-based language and culture required to access the Australian Curriculum. The students could listen and respond in their first language while learning concepts of measurement relevant to the sustainable wild akatyerr harvest context.

2.1 *Linking Anmatyerr and Alyawarr Language and Culture with Mathematics and Science Cross-Curriculum Priorities*

The akatyerr investigation retells the experience of change when grounding pedagogic decision-making in the aspirations of the Utopia Eastern Anmatyerr and Alyawarr community. This section describes the inclusion of Utopia Eastern Anmatyerr and Alyawarr communicators in the curriculum content and pedagogic design. It retells how the Goompi pedagogical cycle (see Chapter 5) was adapted for the akatyerr quadrat study as an intervention in my pedagogical practice. The adjustments to curriculum included creating opportunities for students to interact with local language speakers and in doing so "deepen their knowledge" across the learning areas shown in Table 2 (NTDOE, 2017, p. 3). This built connections between home language and culture and school-based language and culture (NTDOE, 2017). The integrated curriculum included the "language, culture and place of learning" of Utopia Eastern Anmatyerr and Alyawarr speakers (NTDOE, 2017, p. 5). The cultural authority for this area of learning is the "communicative situation" of the Utopia Eastern Anmatyerr and Alyawarr community (ACARA, 2020c).

TABLE 2 Curriculum adjustments for contextualising measurement learning to sustainable wild akatyerr harvest: Integrating Indigenous languages and cultures, science cross-curriculum priorities, and mathematics for Years 8–9 students

Learning area	Indigenous languages and cultures	Science cross-curriculum priorities	Mathematics
Sub-strand	1. Country and land 2. Natural enviroment 3. People and kinship	1. Biology 2. Earth sciences	1. Real number 2. Number and place value 3. Units of measurement
Content description	1. Understand the rules and proper ways to behave on and treat Country at the akatyerr site. 2. Listen to traditional stories about the seasons and how to live on Country; find water and care for Country at the akatyerr site. 3. Understand kinship and Country at the akatyerr site.	1. Investigate Utopia Eastern Anmatyerr and Alyawarr people's responses to environmental threats to important food webs such as akatyerr. 2. Explore Utopia Eastern Anmatyerr and Alyawarr people's connections with, and valuing of, water and water resource management at the akatyerr site.	1. Understand that rate can be solved using percentages – as $\%/m^2$ of akatyerr and other vegetation. 2. Use patterns to assist in finding rules for the multiplication of integers – for amount of vegetation cover inside the quadrat frame. 3. Choosing and using units of area – including cm^2 and m^2 – to measure akatyerr and other vegetation.

(cont.)

TABLE 2 Curriculum adjustments for contextualising measurement learning to sustainable wild akatyerr harvest: Integrating Indigenous languages and cultures, science cross-curriculum priorities, and mathematics for Years 8–9 students (*cont.*)

Learning area	**Indigenous languages and cultures**	**Science cross-curriculum priorities**	**Mathematics**
How the content is explored	Indigenous languages and cultures learning and use	Use and influence of science	Proficiency strands: 1. Fluency 2. Problem-solving 3. Understanding
Communicative situation and texts provided by Utopia Eastern Anmatyerr and Alyawarr speakers.	Listen to oral histories and stories in first language (Utopia Eastern Anmatyerr and Alyawarr). Pool Utopia Eastern Anmatyerr and Alyawarr language knowledge when investigating akatyerr knowledge relating to water, burning, harvesting, and land management.	Record Utopia Eastern Anmatyerr and Alyawarr people's understandings of fire-based agricultural practices for sustainable akatyerr harvest at the akatyerr site.	1. Factorise basic algebraic expression and record in a table as a symbolised pattern. 2. Model real-life vegetation survey situation involving area as a graph drawing (cover of an array) of akatyerr and *Typhonium* plant species (%/m^2). 3. Exchange dialogue to describe pattern in measurement of grass species vegetation (%/m^2).

SOURCE: ADAPTED FROM AUSTRALIAN CURRICULUM, ASSESSMENT AND REPORTING AUTHORITY (2020A, 2020B, 2020C) AND NORTHERN TERRITORY DEPARTMENT OF EDUCATION (2017)

The details from the Australian Curriculum content and processes for exploring content that promoted student access to and agency over contextualised measurement learning are provided in Table 2. The integrated curriculum allowed students to develop fluency in proportional reasoning, for instance, in response to complex real-world situations. Harvest of wild dried akatyerr fruit occurred intermittently (Alyawarr speakers from Ampilatwatja et al., 2009). Real-world problems such as "access to productive lands, narrow economic margins between costs and returns and the skills of younger people" impact the sustainability of the wild harvest and trade of akatyerr (Walsh & Douglas, 2011, p. 412). Improving the skills of younger people and understanding the economic margins in akatyerr trade are complex real-world problems that were able to be addressed through changes to my pedagogic practice.

3 Why Was Intervention in My Pedagogic Practice Necessary?

As pedagogic practices mandated for Northern Territory very remote secondary mathematics education (such as Direct Instruction) are exclusionary, intervention in my practice was necessary. Moreover, Australian Indigenous remote students experience persistent underachievement in numeracy that severely limits their access to further education and life opportunities (ACARA, 2014; Australian Institute of Health and Welfare [AIHW], 2015).

Mathematics education continues to be an area in crisis, particularly for very remote Australian Indigenous students (AIHW, 2021; Rigney, 2011). Furthermore, educational disparity in mathematics is most severely experienced by Indigenous students in very remote settings in the Northern Territory (Australian Government Department of Prime Minister and Cabinet, 2020; Brasche & Harrington, 2012) compared to all other states and territories (Steering Committee for the Review of Government Service Provision [SCRGSP], 2016). Almost one fifth of middle years students (10–14 years old) living in very remote contexts in the Northern Territory are at risk of disengaging entirely from education since they possess limited or no English-speaking skills (SCRGSP, 2016). Additionally, national and international reports of Indigenous students' overall achievement in number, and more broadly mathematics, do not consider male or female students as two distinct groups (OECD, 2017; Thomson et al., 2013). Pedagogic practices that do not consider engendered social norms may contribute to differential educational outcomes for Indigenous students, since social norms influence how male and female students interact (Ganley et al., 2013; Todor, 2014). In addition, a widening gap in science, technology, engineering, and mathematics (STEM) achievement for all female students compared to their male counterparts may be attributed to "limiting teaching practices" (Carrillo, 2019, p. 2).

Remoteness in Indigenous education is associated with the concern of persistent underachievement. My study was situated in a very remote community in the Northern Territory of Australia. Geographic boundaries are often cited when comparing efforts to achieve statistical parity between Indigenous and non-Indigenous students (Australian Government Department of Prime Minister and Cabinet, 2020). The further the distance from the metropolitan, the wider the differences in access to services (Guenther & Osborne, 2020). The preoccupation with achieving the elusive statistical parity (Fogarty et al., 2015) is problematic when attempted through policy measures that continue to delineate differential achievements according to a "geographical construct" based on the metropolitan as the centre (Guenther, Halsey, & Osborne, 2015, p. 75). These measures of success ignore the local community imperatives. That is, "local understandings of purpose, measures of success, and aspiration are ignored" while "metro-centric" measures of success such as "reading, writing and spelling in English" are prioritised (Guenther, Halsey, & Osborne, 2015, p. 67). The preoccupation with the political imperative of achieving statistical parity (Fogarty et al., 2015) diminishes the capacity for local community to develop and lead education grounded in community aspirations and relevant to their "red dirt" context (Guenther et al., 2013). As Rigney et al. (2020, p. 1161) purport, "Australian educational reform has been context-blind" and focused narrowly on teaching strategies to overcome falling and failing assessments in high-stakes tests such as the Programme for International Student Assessment (PISA).

An opportunity to incorporate the unique assets of the learner and the community in which they belonged was presented through the communally held knowledge of sustainable wild harvest practices. The classroom is a unique setting. Scripted curriculum that is delivered in English does not account for and incorporate students' "own distinctive world view" embedded in their first languages (Christie & Harris, 1985, p. 39). For instance, a Yolŋu student will interpret English meaning through the "semantic structures" from their first languages (Christie & Harris, 1985, p. 39). Author and Indigenous education consultant Thelma Perso (2003, p. 10) argues that non-Indigenous teachers should be "explicit in the language they use to teach subject matter" to Indigenous students because of differences in classroom and home language and culture. However, it was not appropriate to deliver Australian measurement curriculum in Utopia Eastern Anmatyerr and Alyawarr classrooms using a scripted and one-size-fits-all didactic approach. Direct Instruction is a model for teaching used in the USA, Canada, UK, and Australia that promises student achievement gains in literacy and numeracy when teachers follow "prescribed instructional practices" that focus on "small learning increments" (National

Institute for Direct Instruction, 2020, "Basic Philosophy" section). However, focusing on building carefully sequenced "skills" rather than emphasising "the whole task" (Perso, 2003, p. 11) may be exclusionary since it fails to respond to the "'ways of knowing' of Aboriginal children" (Harris, 1988, as cited in Perso, 2003, p. 7). Direct Instruction opponents argue that although there is a vast evidence base advocating for "explicit teaching and well-scripted lesson plans" (Ewing, 2011, p. 68), focusing on "small chunks deliberately isolated from the complexities of actual situations" (Hirsch, 2002, p. 63), however, could widen the achievement gap. Hence, my study argued that learning through complex real-world contexts allows students to engage "flexibly and efficiently" with a range of numbers – for instance, "decimals, common fractions, ratio and percent" (Siemon et al., 2005, p. 2). Conversely, the use of Direct Instruction could exacerbate the disadvantage already experienced by a unique cohort of students in a remote context. Despite examples of didactic teaching interactions claiming to improve students' sense of self-efficacy in procedural knowledge (Ewing, 2011; Farkota, 2003; Wilson et al., 2006), at the same time, students may not be given sufficient opportunities to apply their reasoning to new situations.

At the high school where my study was conducted, procedural knowledge – for instance, through worksheets in the Employment Pathways module for area – was recommended for Years 8–9 students (NTDOE, 2021). However, Fogarty et al. (2017) advise that "any program or package that claims to be a one-size-fits-all approach to 'fixing' curriculum and pedagogic practice for Indigenous learners should be treated with caution" (p. 190). It is not appropriate to "claim to know what works" since "what worked" is limited to "the context of a specific moment, community and school context, and learning interaction" (Guenther & Osborne, 2020, p. 2).

A further criticism of scripted lessons is that students are not provided opportunities to construct their own mathematical knowledge. Moreover, when students rather than the pedagogical practices of educators are perceived as the reason for persistent underachievement, educators themselves are not motivated or inclined to reorient their approach to address the mismatch. For instance, ethnomathematics approaches that expect students to "actively construct their own knowledge" have been resisted by mathematics teachers in Zimbabwe (Sunzuma & Maharaj, 2019, p. 13). The teachers tended to resist the change in their teaching role. While positioning students at the centre of learning, the teachers were expected to treat students "as equal partners" in the learning process (Sunzuma & Maharaj, 2019, p. 13). Teachers may have resisted this change as it challenged their perception of who is responsible for teaching and learning.

International research on reform in mathematics demonstrates how the quality of mathematics learning reflects the willingness of educators to transform their pedagogical practices (Wood et al., 2006). According to Wood et al. (2006), the quality of learning when reduced to a series of steps was considered superficial. When teachers resist reform of didactic practices and continue to focus on memorisation of procedural knowledge, this maintains a superficial quality of learning by students. Mathematics seen as a set of steps rather than opportunities to inquire is limiting as it leads to a false perception that there is only one way to solve a mathematical problem.

As discussed earlier, intervention in my pedagogic practice was informed by meetings with women harvesters at bush food forums. Women harvesters were asked by ethno-ecologist Fiona Walsh about what help was needed so they could continue harvesting and selling bush foods (F. Walsh, personal communication, May 30, 2017). The women revealed that they needed support to "maintain seed and fruit production" as well as "more knowledge about how product prices are calculated" (CLC, 2017, p. 16). Students could help to solve this problem, for instance, by applying their proportional reasoning skills to a new circumstance. They calculated a product price for dried akatyerr as dollars per kilogram ($/kg). This used the numeracy skills they had practised when establishing product prices for bush medicine as $/mL and artworks as $/cm^2. Students could then estimate the yield of this important plant as grams per square metre (g/m^2) by reflexively considering lived experiences of collecting akatyerr. Applying this mathematical knowledge to an area of land depicted in Figure 20, students exercised their agency when deciding where they should place 50-metre-long transect lines (and later estimate the plant distribution across a rhombus; represented in Figures 47 and 48). This caring-for-country strategy of monitoring plant species is a method for protecting akatyerr seeds and fruit contained within the site surveyed.

Plant distribution, or the amount of vegetation cover inside a quadrat frame of 1 m^2 (100 cm × 100 cm), was estimated by exploring the connection between number, place value, and measurement. Students could communicate their solutions, from the factorisation of basic algebra for instance, in ways that were relevant to the context. Students constructed real-life representations of the proportion of the quadrat frame area covered in akatyerr and other vegetation observed. The contextualised curriculum was designed from the context of the learner, which ensured the students' values, beliefs, histories, knowledges, stories, and cultural practices were valued (Moggridge & Mihinui, 2020). The learner could freely express who they are while constructing new mathematics learning. Students were able to access mathematics education as articulated in the Australian Curriculum in a manner that was culturally safe (AITSL, 2020). That is, students were able to reinforce their identity, their cultural traditions, and their belief in themselves as capable mathematics learners, while

FIGURE 20 A student points to the direction the transect line will be laid
PHOTOGRAPH BY TIM WARE

conducting the akatyerr investigation (Fogarty et al., 2015; Matthews, 2012; Sarra, 2011). The design of socially just and equitable mathematics pedagogy enabled students to feel that who they are was valued and integrated with – not displaced by – other worldviews (Lingard, 2005). Developing proficiency with proportional reasoning is particularly important for middle years students (Lamon, 2012) and is necessary for learning measurement (Dole et al., 2012). Sustainable wild akatyerr harvest knowledge concepts were familiar to students, assistant teachers, and community members. Students were supported to learn measurement from a strengths discourse by drawing from their lived experiences and interests in harvesting akatyerr.

The following chapter reveals the decolonising actions that led to traditional ecological knowledge – that is, a "cumulative body of knowledge, practice and belief, evolving by adaptive processes" (Janke, 2021, p. 389) – being shared with the younger generation in yarning circles.

References

Alfred, T. (2009). Colonialism and state dependency. *Journal of Aboriginal Health, 5*, 42–60. Retrieved January 11, 2023, from https://jps.library.utoronto.ca/index.php/ijih/article/view/28982

Altman, J. (2009). *Beyond closing the gap: Valuing diversity in Indigenous Australia* (Working Paper No. 54). Centre for Aboriginal Economic Policy Research. Retrieved November 9, 2022, from https://openresearch-repository.anu.edu.au/handle/1885/147808

Alyawarr speakers from Ampilatwatja, Walsh, F., & Douglas, J. (2009). *Angka Akatyerr-akert: A desert raisin report.* Desert Knowledge Cooperative Research Centre. Retrieved January 11, 2023, from https://www.nintione.com.au/resource/DKCRC_Angka-Akatyerr-akert_A-Desert-raisin-report.pdf

Australian Curriculum, Assessment and Reporting Authority. (2014). *National Assessment Program – Literacy and Numeracy: Achievement in reading, persuasive writing, language conventions and numeracy: National report for 2014.* https://nap.edu.au/_resources/National_Assessment_Program_Literacy_and_Numeracy_national_report_for_2014.pdf

Australian Curriculum, Assessment and Reporting Authority. (2020a). *Australian curriculum cross-curriculum priorities: Aboriginal and Torres Strait Islander histories and cultures.* Retrieved January 11, 2023, from https://australiancurriculum.edu.au/f-10-curriculum/cross-curriculum-priorities/aboriginal-and-torres-strait-islander-histories-and-cultures/

Australian Curriculum, Assessment and Reporting Authority. (2020b). *Australian curriculum: Learning continuum of numeracy: Using fractions, decimals, percentages, ratios and rates (Level 5).* Retrieved January 11, 2023, from https://bit.ly/3GzjB4i

Australian Curriculum, Assessment and Reporting Authority. (2020c). *Australian curriculum: Mathematics (Version 8.4) – Year 8 level description.* Retrieved January 11, 2023, from https://bit.ly/3GB8zM7

Australian Curriculum, Assessment and Reporting Authority. (2020d). *Australian curriculum: Mathematics (version 8.4) – Year 8 content descriptions: Number and algebra.* Retrieved January 11, 2023, from https://bit.ly/3Xkbv6v

Australian Curriculum, Assessment and Reporting Authority. (2020e). *National numeracy learning progression – Comparing units (ratios, rates and proportion) description.* Retrieved January 11, 2023, from https://www.australiancurriculum.edu.au/resources/national-literacy-and-numeracy-learning-progressions/national-numeracy-learning-progression/number-sense-and-algebra/?subElementId=50723&scaleId=0ssssw

Australian Government Department of Prime Minister and Cabinet. (2020). *Closing the gap report 2020.* Commonwealth of Australia. Retrieved January 11, 2023, from https://ctgreport.niaa.gov.au/sites/default/files/pdf/closing-the-gap-report-2020.pdf

Australian Institute of Health and Welfare. (2015). *The health and welfare of Australia's Aboriginal and Torres Strait Islander peoples: 2015.* Australian Government. https://doi.org/10.25816/5ebcbd26fa7e4

Australian Institute of Health and Welfare. (2021, September 16). *Secondary education: School retention and completion.* Australian Government. Retrieved January 25, 2023, from https://www.aihw.gov.au/reports/australias-welfare/secondary-education-school-retention-completion

Australian Institute of Teaching and School Leadership. (2020). *Indigenous cultural competency in the Australian teaching workforce: Discussion paper.* Retrieved January 11, 2023, from https://www.aitsl.edu.au/docs/default-source/comms/cultural-competency/aitsl_indigenous-cultural-competency_discussion-paper_2020.pdf

Australian Native Food and Botanicals. (n.d.). *Growing the grower: Farm co-operatives and collaboration roadshow.* Retrieved November 8, 2022, from https://anfab.org.au/main.asp?_=Growing%20the%20Grower

Boyd, N. (2017, August). *Akatyerr: Career pathways on akatyerr* [PowerPoint slides]. Northern Territory Department of Education.

Brasche, I., & Harrington, I. (2012). Promoting teacher quality and continuity: Tackling the disadvantages of remote Indigenous schools in the Northern Territory. *Australian Journal of Education, 56*(2), 110–125. https://doi.org/10.1177/000494411205600202

Carrillo, J. (2019). *Mindset, math and gender: How impacting female students' mindset towards math impacts their success in the classroom and beyond* [Doctoral dissertation, California State University, Fullerton]. ProQuest Dissertations Publishing. Retrieved January 11, 2023, from https://www.proquest.com/openview/0851aa7678f8d47d227f7081d7c34bed/1?pq-origsite=gscholar&cbl=2026366&diss=y

CAYLUS (Central Australian Youth Link Up Service). (2017). *Stronger families for communities* [Unpublished report].

Central Land Council. (2017, August). Seed collectors want a hand to meet growing bush foods demand. *Land Rights News Central Australia, 7*(2), 16. Retrieved December 9, 2022, from https://www.clc.org.au/land-rights-news/

Christie, M., & Harris, S. (1985). Communication breakdown in the Aboriginal classroom. In J. B. Pride (Ed.), *Cross-cultural encounters: Communication breakdown and mis-communication.* River Scene.

Cornell, S. (2006). *Indigenous peoples, poverty and self-determination in Australia, New Zealand, Canada and the United States* (Joint Occasional Papers on Native Affairs, JOPNA No. 2006-02). Native Nations Institute for Leadership, Management, and Policy and Harvard Project on American Indian Economic Development. Retrieved January 10, 2023, from https://www.researchgate.net/publication/238707334_Indigenous_Peoples_Poverty_and_Self-Determination_in_Australia_New_Zealand_Canada_and_the_United_States

Corntassel, J. (2012). Re-envisioning resurgence: Indigenous pathways to decolonization and sustainable self-determination. *Decolonization: Indigeneity, Education & Society, 1*(1), 86–101. Retrieved January 11, 2023, from https://jps.library.utoronto.ca/index.php/des/article/view/18627

Desart (Association of Central Australian Aboriginal Arts and Crafts Centres). (2018). *Desert Mob 2018 Symposium: Celebrating 25 years of Desart.* Retrieved November 8, 2022, from https://desart.com.au/wp-content/uploads/sites/40/Desert-Mob-Symposium-2018-Digital.pdf

Dole, S., Clarke, D., Wright, T., & Hilton, G. (2012). Students' proportional reasoning in mathematics and science. In T. Y. Tso (Ed.), *Proceedings of the 36th conference of the International Group for the Psychology of Mathematics Education* (pp. 195–202). PME.

Ewing, B. (2011). Direct instruction in mathematics: Issues for schools with high Indigenous enrolments: A literature review. *Australian Journal of Teacher Education, 36*(5), 65–92. https://doi.org/10.14221/ajte.2011v36n5.5

Farkota, R. (2003). *The effects of a 15-minute Direct Instruction intervention in the regular mathematics class on students' mathematics self-efficacy and achievement* [Doctoral thesis, Monash University]. Australian Council for Educational Research. Retrieved January 11, 2023, from http://www.acer.edu.au/documents/FarkotaThesis.pdf

Fogarty, W., Lovell, M., & Dodson, M. (2015). A view beyond review: Challenging assumptions in Indigenous education development. *UNESCO Observatory Multi-Disciplinary Journal in the Arts, 4*(2), 1–23. Retrieved October 14, 2022, from https://hdl.handle.net/1885/13420

Fogarty, W., Riddle, S., Lovell, M., & Wilson, B. (2017). Indigenous education and literacy policy in Australia: Bringing learning back to the debate. *Australian Journal of Indigenous Education, 47*(2), 185–197. https://doi.org/10.1017/jie.2017.18

Ganley, C. M., Mingle, L. A., Ryan, A. M., Ryan, K., Vasilyeva, M., & Perry, M. (2013). An examination of stereotype threat effects on girls' mathematics performance. *Developmental Psychology, 49*(10), 1886–1897. https://doi.org/10.1037/a0031412

Green, J., Blackman, D., & Moore, D. (2019). *Alyawarr to English Dictionary* (2nd ed.). IAD Press.

Guenther, J., Bat, M., & Osborne, S. (2013). Red dirt thinking on educational disadvantage. *The Australian Journal of Indigenous Education, 42*(2), 100–110. https://doi.org/10.1017/jie.2013.18

Guenther, J., Halsey, J., & Osborne, S. (2015). From paradise to beyond: Geographical constructs and how they shape education in the bush. *Australian and International Journal of Rural Education, 25*(3), 62–79. Retrieved January 25, 2023, from https://journal.spera.asn.au/index.php/AIJRE/article/view/104/78

Guenther, J., & Osborne, S. (2020). Did DI do it? The impact of a program designed to improve literacy for Aboriginal and Torres Strait Islander students in remote schools. *The Australian Journal of Indigenous Education, 49*(2), 163–170. https://doi.org/10.1017/jie.2019.28

Hirsch, E. D. J. (2002). Classroom research and cargo cults. *Policy Review, 115*, 51–70.

Janke, T. (2016, December 21). *True tracks: Create a culture of innovation with Indigenous knowledge* [Video]. YouTube. Retrieved December 8, 2022, from https://www.youtube.com/watch?v=WfS11_Dl6ew

Janke, T. (2021). *True tracks: Respecting Indigenous knowledge and culture.* NewSouth Publishing.

Lamon, S. (2012). *Teaching fractions and ratios for understanding: Essential content knowledge and instructional strategies.* Taylor & Francis Group.

Lingard, B. (2005). Socially just pedagogies in changing times. *International Studies in Sociology of Education, 15*(2), 165–186. https://doi.org/10.1080/09620210500200138

Lingard, B., & Martin, P. (2016). Strategies to support the interests of Aboriginal and Torres Strait Islander peoples in the commercial development of gourmet bush food products. *International Journal of Cultural Property, 23*(1), 33–70. https://doi.org/10.1017/S0940739116000023

Logue, D., Pitsis, A., Pearce, S., & Chelliah, J. (2018). Social enterprise to social value chain: Indigenous entrepreneurship transforming the native food industry in Australia. *Journal of Management & Organization, 24*(2), 312–328. https://doi.org/10.1017/jmo.2017.24

Maclean, K., Woodward, E., Jarvis, D., Rowland, D., Rist, P., Turpin, G., Martin, P., & Glover, R. (2019). *A strategic sector development and research priority framework for the traditional owner-led bush products sector in northern Australia.* CSIRO. https://doi.org/10.25919/5d72acc375ac7

Matthews, C. (2012). Maths as storytelling: Maths is beautiful. In K. Price (Ed.), *Aboriginal and Torres Strait Islander education: An introduction for the teaching profession* (pp. 94–112). Cambridge University Press.

McRae-Williams, E. (2014). Aspirations for meaningful livelihoods: Challenges of pathway navigation. *Journal of Australian Indigenous Issues, 17*(4), 57–71. Retrieved November 9, 2022, from https://www.worldcat.org/title/926653601

Merne Altyerre-ipenhe [Food from the Creation time] Reference Group, Douglas, J., & Walsh, F. (2011). *Aboriginal people, bush foods knowledge and products from central Australia: Ethical guidelines for commercial bush food research, industry and enterprises* (Desert Knowledge Cooperative Research Centre Report No. 71). Ninti One Limited. Retrieved January 11, 2023, from https://www.nintione.com.au/resource/NintiOneResearchReport_71_BushFoodGuidelines.pdf

Moggridge, B., & Mihinui, R. (2020). *Australian & New Zealand guidelines for fresh & marine water quality: Indigenous principles for water quality.* Water Quality Australia. Retrieved January 11, 2023, from https://www.waterquality.gov.au/anz-guidelines/guideline-values/derive/cultural-values/principles

National Institute for Direct Instruction. (2020). *Basic philosophy of Direct Instruction (DI).* Retrieved January 11, 2023, from https://www.nifdi.org/what-is-di/basic-philosophy.html

Northern Territory Board of Studies. (2017). *Keeping Indigenous languages and culture strong: A plan for the teaching and learning of Indigenous languages and cultures in Northern Territory Schools*. Northern Territory Government. Retrieved January 11, 2023, from https://education.nt.gov.au/__data/assets/pdf_file/0012/413202/Policy_Keeping-Indigenous-Languages-and-Cultures-Strong-Document_web_updated.pdf

Northern Territory Department of Education. (2016a). *Indigenous education strategy employment pathways: Country, culture, careers: Big business – Running an enterprise* (*Version 3.1*). Northern Territory Government.

Northern Territory Department of Education. (2016b, January 22). *Indigenous education strategy newsletter – Issue 1*. Northern Territory Government. Retrieved January 11, 2023, from https://education.nt.gov.au/statistics-research-and-strategies/indigenous-education-strategy/indigenous-education-strategy-newsletter-issue-1

Northern Territory Department of Education. (2016c). *Northern Territory employment pathways – Teacher handbook* (*Version 0.0.5*). Northern Territory Government.

Northern Territory Department of Education. (2017). *NT Indigenous languages and cultures: First Languages Maintenance (L1M)*. Northern Territory Government. Retrieved January 11, 2023, from https://education.nt.gov.au/__data/assets/pdf_file/0020/715232/ntilc-l1m-nov18-03.pdf

Northern Territory Department of Education. (2021). *Indigenous education strategy employment pathways: Communication and problem solving: Numeracy – Area* (*Version 2.2*). Northern Territory Government. Retrieved January 11, 2023, from http://ed.ntschools.net/partpath/team/epprograms/SiteAssets/SitePages/Communication%20and%20Problem%20Solving/Areav2.2.pdf

Northern Territory Department of Environment, Parks and Water Security. (2021). *Threatened species of the Northern Territory: Sandover lilly – Typhonium sp.* Sandover. Retrieved November 8, 2022, from https://nt.gov.au/__data/assets/pdf_file/0018/208503/typhonium-sp-sandover.pdf

Northern Territory Government. (2017, July 19). *Remote secondary school choices*. Retrieved November 9, 2022, from https://nt.gov.au/learning/primary-and-secondary-students/remote-students-and-parents/remote-secondary-school-choices

Organisation for Economic Co-operation and Development. (2017). *2013 OECD recommendation of the council on gender equality in education, employment and entrepreneurship*. OECD Publishing. https://doi.org/10.1787/9789264279391-en

Perso, T. (2003). *Improving Aboriginal numeracy: A book for education systems, school administrators, teachers and teacher educators*. Mathematics, Science & Technology Education Centre (MASTEC), Edith Cowan University. Retrieved January 11, 2023, from https://qut.primo.exlibrisgroup.com/permalink/61QUT_INST/1g7tbfa/alma991009944788704001

Reconciliation Australia. (2019, February 7). *RA IGA S 2018, category B finalist, Apmer Akely-akely for the SCFC Utopia project* [Video]. YouTube. Retrieved October 12, 2022, from https://www.youtube.com/watch?v=9k3qwfvgkZY

Rigney, L. I. (2011). Indigenous education and tomorrow's classroom: Three questions, three answers. In N. Purdie, G. Milgate, & H. R. Bell (Eds.), *Two way teaching and learning: Toward culturally reflective and relevant education* (pp. 35–47). Allen & Unwin.

Rigney, L. I., Garrett, R., Curry, M., & MacGill, M. (2020). Culturally responsive pedagogy and mathematics through creative and body-based learning: Urban Aboriginal schooling. *Education and Urban Society*, *52*(8), 1159–1180. https://doi.org/10.1177/0013124519896861

Sarra, C. (2011). *Strong and smart – Towards a pedagogy for emancipation: Education for first peoples*. Routledge.

Siemon, D., Breed, M., & Virgona, J. (2005). From additive to multiplicative thinking: The big challenge of the middle years. In J. Mousley, L Bragg, & C. Campbell (Eds.), *Mathematics – celebrating achievement: Proceedings of the 42nd conference of the Mathematical Association of Victoria*. MAV.

Steering Committee for the Review of Government Service Provision. (2016). *Overcoming Indigenous disadvantage: Key indicators 2016*. Productivity Commission. Retrieved January 25, 2023, from https://www.pc.gov.au/research/ongoing/overcoming-indigenous-disadvantage/2016

Sunzuma, G., & Maharaj, A. (2019). Teacher-related challenges affecting the integration of ethnomathematics approaches into the teaching of geometry. *EURASIA Journal of Mathematics, Science and Technology Education*, *15*(9), Article em1744. https://doi.org/10.29333/ejmste/108457

Thomson, S., De Bortoli, L., & Buckley, S. (2013). *PISA 2012: How Australia measures up: The PISA 2012 assessment of students' mathematical, scientific and reading literacy*. Australian Council for Educational Research. Retrieved January 11, 2023, from https://research.acer.edu.au/ozpisa/15

Todor, I. (2014). Investigating "the old stereotype" about boys/girls and mathematics: Gender differences in implicit theory of intelligence and mathematics self-efficacy beliefs. *Procedia – Social and Behavioral Sciences*, *159*, 319–323. https://doi.org/10.1016/j.sbspro.2014.12.380

Walsh, F., & Douglas, J. (2011). No bush foods without people: The essential human dimension to the sustainability of trade in native plant products from desert Australia. *The Rangeland Journal*, *33*(4), 395–416. https://doi.org/10.1071/RJ11028

Watson, I. (2018). Aboriginal recognition: Treaties and colonial constitutions, "we have been here forever ...". *Bond Law Review*, *30*(1), 7–18. https://doi.org/10.53300/001c.5657

Wilson, B. (2014). *A share in the future. Review of Indigenous education in the Northern Territory.* The Education Business. Retrieved January 11, 2023, from https://www.nt.gov.au/__data/assets/pdf_file/0020/229016/A-Share-in-the-Future-The-Review-of-Indigenous-Education-in-the-Northern-Territory.pdf

Wilson, B., Abbott, T., Quinn, S. J., Guenther, J., McRae-Williams, E., & Cairney, S. (2019). Empowerment is the basis for improving education and employment outcomes for Aboriginal people in remote Australia. *The Australian Journal of Indigenous Education, 48*(2), 153–161. https://doi.org/10.1017/jie.2018.2

Wilson, L., Andrew, C., & Below, J. (2006). A comparison of teacher/pupil interaction within mathematics lessons in St Petersburg, Russia and the north-east of England. *British Educational Research Journal, 32*(3), 411–441. Retrieved January 11, 2023, from https://www.jstor.org/stable/30032676

Wood, T., Shin, S., & Doan, P. (2006). Mathematics education reform in three US classes. In D. Clarke, C., Keitel, & Y. Shimizu (Eds.), *Mathematics classrooms in twelve countries: The insider's perspective* (pp. 75–86). Sense Publishers. https://doi.org/10.1163/9789087901622_006

CHAPTER 4

Preferencing Utopia Eastern Anmatyerr and Alyawarr Voice in Pedagogy

The Northern Territory First Nations have a deep record of sovereignty, which is now recognised in *The Barunga Agreement 8 June 2018: A Memorandum of Understanding* (Aboriginal Land Councils and NTG, 2018):

> It is agreed that:
> a. Aboriginal people, the First Nations, were the prior owners and occupiers of the land, seas and waters that are now called the Northern Territory of Australia.
> b. The First Nations of the Northern Territory were self-governing in accordance with their traditional laws and customs; and that
> c. First Nations peoples of the Northern Territory never ceded sovereignty of their lands, seas and waters. (p. 6)

Further, in the Northern Territory, the aspirations of Makarrata – "the coming together after a struggle" – are embedded in this agreement (Referendum Council, 2017, para. 10). That is, the "aspirations for a fair and truthful relationship with the people of Australia" (Referendum Council, 2017, para. 10) are captured in the processes for agreement-making. To reconcile past injustices and provide "a better future for our children based on justice and self-determination" (Referendum Council, 2017, para. 10), the Barunga Agreement truthfully tells that

> there has been deep injustice done to the Aboriginal people of the Northern Territory, including violent dispossession, the repression of their languages and cultures, and the forcible removal of children from their families, which have left a legacy of trauma, and loss that needs to be addressed and healed. (Aboriginal Land Councils and NTG, 2018, p. 6)

In Aotearoa (New Zealand), a teacher's responsiveness to the Treaty of Waitangi – a signed agreement "promising partnership between Māori and the British crown and protection of Māori taonga (treasures)" – is embedded in the New Zealand Teacher Professional Standards (Averill et al., 2020, p. 114). In

 | DOI:10.1163/9789004682757_004

Australia, the Commonwealth Government has never established a treaty with our First Nations peoples. The principles guiding a Treaty consultative process in the Northern Territory as stated in the Barunga Agreement are between the Aboriginal Land Councils (Northern Land Council, CLC, Anindilyakwa Land Council, and Tiwi Land Council) and the NTG. Also guiding the Treaty agreement-making processes are Articles of the UNDRIP (United Nations, 2007). As discussed in Chapter 1, Section 2.3, UNDRIP recognises cultural heritage rights such as the right for Indigenous peoples to "maintain, control, protect and develop their cultural heritage, traditional knowledge and traditional cultural expressions ... including ... seeds, medicines, knowledge of the properties of fauna and flora, oral traditions ..." (United Nations, 2007, Article 31, para. 1).

As a non-Indigenous woman engaged in transforming pedagogy, I supported the right for the students "to be educated in [their] own languages and in [their] own culture and history" (Barunga Statement, 1988, as cited in Aboriginal Land Councils and NTG, 2018). To do this, I engaged in a space with Eastern Anmatyerr and Alyawarr peoples and promoted their voices "above" my own (Finlay, 2020). Yorta Yorta woman Summer May Finlay, speaking on her experiences with non-Indigenous allies, suggests that preferencing Aboriginal voices involves "supporting us, working with us, but allowing us to speak on the subjects that affect our lives. If you're really here for us, you won't mind taking a step back" (Finlay, 2021, "Preference our voices", para. 3). Elevating the voices of mob participating in this study occurred during a prolonged engagement and "in depth pursuit" of the ways to contextualise measurement learning to the traditional practices involved in wild akatyerr fruit harvest (Schwandt et al., 2007, p. 18). Ongoing consultation in yarning circles (see Figure 21) ensured the voices of the Utopia Eastern Anmatyerr and Alyawarr community guided the decision-making processes. I worked alongside mob by allowing them to speak while I took a step back to listen, self-reflect, and consider how to adjust my pedagogic practice.

The students depicted in Figure 21 were actively listening and responding to the right arrangement of people in relationally informed ways (Christie et al., 2010). Pedagogical practices can recognise the "relational, place-based existence" of First Nations students (Corntassel, 2012, p. 13). Listening and adjusting pedagogy with respect to relationality supports First Nations rights to self-determination. Pedagogical adjustments, such as the arrangement of people, provided students in yarning circles "the freedom to pursue [their] own economic, social, religious and cultural development" (Barunga Statement, 1988, as cited in Aboriginal Land Councils and NTG, 2018). Yarning, listening, critically reflecting, and assessing plant distribution integrated worldviews in measurement learning.

FIGURE 21 A yarning circle conducted on the akatyerr growing site, showing the relational processes. Two elders, Jedda Purvis (blue beanie) and Judy Purvis (pink beanie), with the assistant teacher, Louisa Long (red shirt), and the teacher, Nicole Boyd (blue shirt), are interacting during the student-led inquiry
PHOTOGRAPH BY TIM WARE

1 Creating the Space and Place for Merging Epistemologies in Measurement Curriculum

The space for merging worldviews in curriculum is at the site of knowledge construction. In the school context, the places where knowledge is constructed are the sites of the coloniser's oppression and power – and the place to counter colonialism (Rigney, 1999). The emphasis on listening to and preferencing the community concerns regarding knowledge sharing on akatyerr supported community control over cultural heritage knowledge (see UNDRIP, United Nations, 2007, Article 31, para. 1). Community control relating to "traditional acts and culture" in the study occurred *as* and *when* it was embedded in the measurement curriculum at the akatyerr harvest site (Janke & Cubillo, 2019, p. 1). The following sections retell the yarning circle dialogue that allowed this to occur. Stepping back and listening to Utopia Eastern Anmatyerr and Alyawarr harvesters was an adjustment I made in my pedagogic practice. This allowed the right people to speak for and on country. "Having the opportunity to go out bush to where our language originated" is particularly important for mob who have few remaining speakers – as it forms an "important part of our culture and a strong foundation of our identities" (CLC, 2018, p. 9).

1.1 *Who Are the* Right People *to Speak About Akatyerr?*

The people interested in protecting akatyerr harvest interests joined initial yarning circles and discussed kinship relations among harvesters (elders, young women), traditional land managers (men), and me. Written and verbal interactions among each of us in yarning circles provided opportunities to consider a suitable blend of epistemologies for the students' investigation of akatyerr based on concepts of interest to them (see Figure 12; Dobson et al., 2008; Merne Altyerr-ipenhe Reference Group et al., 2011; Turner, 2005). Oral and written dialogue captured the interactions and, on critical reflection, allowed me to challenge my assumptions of what "correct rules and ways of proceeding" look like when planning measurement learning (Scherer & Steinbring, 2006, p. 174). In an intercultural space, planning reflected the "relationship between a person and that country" – which is known according to kin terms (Green et al., 2019, p. 423). Lucy Kunoth is an Eastern Anmatyerr and Alyawarr woman who participated in the study. As a community member, she shared intergenerational knowledge relating to bush foods and bush trips (or country visits). For Lucy, the country chosen was based on her relationship to her mother or "opposite generation moiety" (Green et al., 2019, p. 426). She chose that country based on her mother's "relationships to land" (Green et al., 2019, p. 425). Kinship relationships also determined the obligation for her to teach the next generation about her mother's country. When I invited Lucy to participate in the first yarning circle meeting, she opened the yarning circle meeting with the following introduction:

> Hello. My name is Lucy. I come from the outstation. That's where my mother's country is. And I'd like to take some of the school kids on a trip for bush tucker, akatyerr, and I know a good place is my mother's country, where a lot of akatyerr grow. That's a good place we can go. And bush trip and tell more stories and show my mother's country and tell some more stories. And there's a lot of bush medicines grows there, and we can go have a look around there ... And you can walk around and look for akatyerr. And if there's some ready, we can pick some ... I can see already that green one starting from rain. When we eat 'em that yellow one might be ready. But it's good to get drier one. Summertime they dry 'em up. Summertime [pause] And it's really good to eat yellow one. Sweet one ... Get bucket, big jar, and fill 'em up. I know where that place is, not far. Close. (Yarning circle SS1, May 16, 2019)

Lucy shared some of the akatyerr harvest protocols in this meeting. She knew where to go and which fruit to harvest. Lucy identified that she was interested in taking students on a bush trip to learn about akatyerr on her mother's country. From Lucy's experience, this akatyerr site was considered "a good place we can go" since it was "where a lot of akatyerr grow". Lucy identified her family relationships to the chosen site and her interest and obligations to the site through this kinship relationship. As documented by Green et al. (2019), "kinship is a defining part of social and cultural life" (p. 422). Lucy had considered what the ways of proceeding would look like. I sat back and listened and adjusted my practice to reflect her suggestions as to where and how the akatyerr investigation would occur.

Later, assistant teachers Louisa and Sharmic, who were also related by kin to the site, guided the decision-making around how the measurement learning would be contextualised to akatyerr harvest knowledge concepts related to people as and when the learning occurred (Dobson et al., 2008; Merne Altyerr-ipenhe Reference Group et al., 2011; Turner, 2005). Louisa and Sharmic's voices were critical to ensure rules relating to kinship were closely observed while conducting the investigation. Knowledgeable in their kinship system Louisa and Sharmic responded to the "relationships between people ... and their traditional country, the plants and animals that originate in that country" (Green et al., 2019, p. 422). The women arranged the suitable kin to participate in the study at the agreed akatyerr site. The seating of participants within yarning circles was arranged by the adults to respect the kinship between people and in relation to the traditional country – the akatyerr site.

1.2 *Who Are the* Right People *to Harvest Akatyerr?*

Lucy and I ensured that the *right people* were invited to visit the akatyerr site. Decision-making regarding who should speak about akatyerr knowledge concepts of interest to students was shared among Utopia Eastern Anmatyerr and Alyawarr adults related by kinship to the site. Lucy, in the following excerpt, determined that the student participants were able to visit the site for harvesting akatyerr:

> Nicole: Is it okay for girls [yarning circle Years 8–9 girls] to pick from your country?
>
> Lucy: It's good [spoken softly]. Yeah. This mob [yarning circle girls] can pick ... Anyone. Just go together [spoken softly]. Not wander around. Just stick together ... Oh yeah. We're all family. We're [emphasised] *allowed* to go. (Yarning circle SS1, May 16, 2019)

This dialogue exchanged at a school site (SS) developed the concept of appropriate people. Lucy asserted her cultural authority on wild harvest protocols by stating: "We're all family. We're *allowed* to go".

1.3 *What Are the* Right Ways *to Blend Epistemologies?*

I was aware of the measurement attributes that could be contextualised with akatyerr harvest knowledge concepts from measurement curriculum recommendations presented by Alyawarr speakers from Ampilatwatja et al. (2009, pp. 48–49). I initiated a conversation with Lucy to identify the appropriateness of measurement attributes such as *area* being applied to the akatyerr site chosen to visit. The following excerpt from my field notes retells the interaction I had with Lucy during the first yarning circle:

> I just listened to [Lucy] and didn't interrupt her and then after a while I prompted with a couple of other questions to do with size [of the site], whether we could walk around there ... (Nicole Boyd, field notes, May 16, 2019)

Listening to Lucy, I thought about whether length, perimeter, and area could be explored at the site. I asked whether the participants could walk around the akatyerr site to deepen their understanding of the measurement attributes under investigation. In the following excerpt from the first yarning circle, Lucy determined that we could walk around and in doing so she exercised the right to "express and strengthen" connections to Utopia lands (Janke, 2021, p. 236).

> Nicole: Is it a big area? Can you walk around it?
> Lucy: And you can walk around and look for akatyerr. And if there's some ready, we can pick some ... It goes [grows] everywhere. (Yarning circle SS1, May 16, 2019)

As a non-Indigenous Australian, reflecting on Rigney's guidelines for Indigenist research, I asked myself whether I could "inform the struggles of Indigenous Australians for genuine self-determination" (1999, p. 117). I knew that measurement attributes such as area could be contextualised to cultural production such as original artworks (see Chapter 3, Section 1.5). I conceptualised that land could be viewed in a similar way to a canvas – as a continuous surface that could be divided into discrete unit of measurement. However, only Utopia Eastern Anmatyerr and Alyawarr students and adults participating in the investigation were able to "set their own agenda for liberation" (Rigney, 1999, p. 117) with respect to sharing knowledge on the cultural production of bush foods like akatyerr. As Yorta Yorta academic Summer May Finlay argues, genuine self-determination supports mob "to speak on the subjects that affect [their] lives" (2021, "Preference our voices", para. 3). This meant listening to and responding to the directions given by mob while planning and delivering lessons.

1.4 *What Are the* Right Ways *to Learn From Country?*

Lucy shared essential information in the first yarning circle she participated in. The dialogue exchanged informed the ways to act. The information shared developed the protocols expected while visiting her mother's country:

> Lucy: [spoken softly] Just go together. Not wander around. Just stick together. (Yarning circle SS1, May 16, 2019)

Listening to Lucy's direction was vital to protect the knowledge of akatyerr with respect to the people and their place while on country. Participants may have accidentally visited the wrong place. There may have been sites, for instance, that female students were not permitted to visit according to kinship considerations.

1.5 *How Do You Protect Secret Knowledge Shared in Yarning Circles?*

Dialogue exchanged in yarning circles was controlled by Utopia Eastern Anmatyerr and Alyawarr participants. This included who spoke, what was said, and how it was said (in first languages and English language). For instance, Louisa spoke in Anmatyerr to respond to the questions asked in English. For each question, Louisa prompted the student leading the inquiry to indicate when it was time to ask the next question. Later, the translation and transcription of the two-way dialogue on akatyerr was checked and controlled by the Utopia Eastern Anmatyerr and Alyawarr participants so "secret or sacred knowledge" was protected (Janke & Cubillo, 2019, p. 2). The ICIP principle that relates to "secrecy and privacy" was applied; for instance, some of the transcription data were redacted so that the right for Indigenous people to self-determination was also protected (Janke & Cubillo, 2019, p. 2).

2 Decolonising Mathematics Education

Preferencing First Nations' voices in relation to the students' real-world-context interests expanded the space for Indigenous pedagogies (epistemologies) to sit alongside Western pedagogies. Strengthening student identity through first language and culture content provided an interface for students to access measurement curriculum as and when it intersected with akatyerr knowledge concepts. Critically reflecting on the themes that emerged from the translated and transcribed yarning circle dialogue (see Figure 22) supported ongoing acts of sovereignty. The Utopia Eastern Anmatyerr and Alyawarr people participating in the study decided if these themes accurately represented what was said in the yarning circles.

2.1 *Emerging Themes*

Thematic analysis generated a broad understanding from repeated patterns (themes) that emerged across the entire data set (Braun & Clarke, 2006). That is, the way I contextualised students' interests and lived experiences of harvesting akatyerr emerged while familiarising myself with all field data, assigning codes, and analysing all possible themes. Describing the procedures for transcribing and coding data strengthened the validity of the patterns observed (Strauss, 1987). These details also added rigour to the coding and analysis effort (Strauss, 1987, p. 27). There were no known codes previous to the study for the phenomenon under investigation. Coding for the specific research question evolved during the coding process.

The initial thematic analysis revealed nine potential themes as depicted in Figure 22. NVivo 12 Plus software was used to generate these themes. This

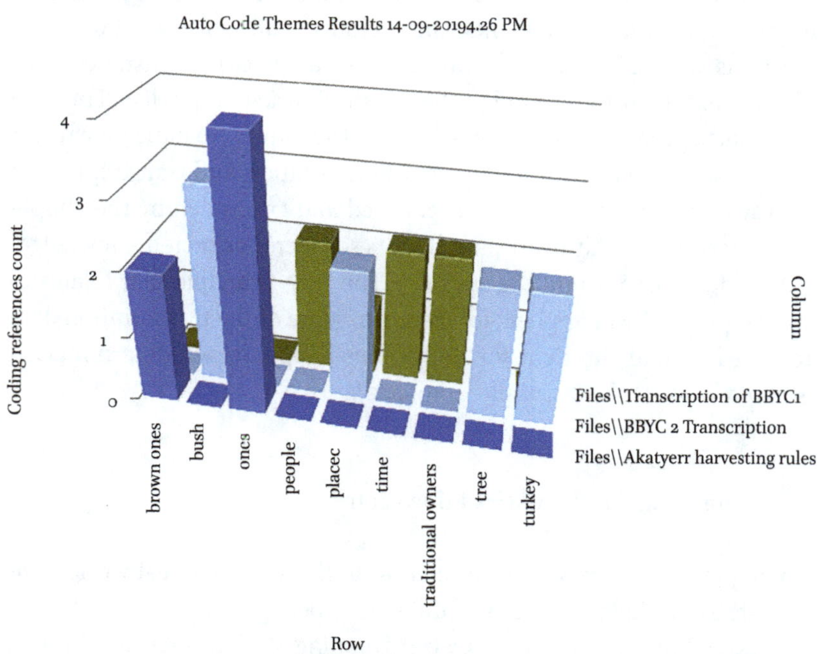

FIGURE 22 Auto-coded themes generated from yarning circles at the akatyerr site

software was used initially to check the preliminary word frequency of key words and phrases naturally occurring across all transcripts. Similar codes that were aggregated in the database, such as young people, family, traditional owners, and groups, suggested possible themes since together the codes contributed to a major idea, such as people. Student artefacts were later selected to show where students made connections to big pedagogical ideas critical to their measurement learning. A reflection-in-action approach (Schon, 1983) enabled me to select student artefacts that revealed the use of reflexive strategies such as generalisation, for instance, in their multiple representations (YuMi Deadly Centre, 2014, pp. 19–20). The description of the teaching and learning experiences was derived mostly from the visit to an important akatyerr growing site. Extracts of the interactions between Utopia Eastern Anmatyerr and Alyawarr adults, the students, and me were used to strengthen the interpretation of the patterns observed.

When defining and naming themes, the assistant teachers, critical friend, and I considered the audiovisual data according to who spoke, what was said, and how it was said (for instance, indirectly in creation stories). By checking with participants the accuracy of the transcribed data and checking the permission for the text to be read by a wider audience, we continued to negotiate and respect our agreement made at informed consent. That is, that we would act in a way to maintain cultural protocols while supporting students to learn intergenerational knowledge at the Akatyerr site. Member checking occurred in discussion with the adult participants. Member checking enabled the participants to check the accuracy of re-storied dialogue to protect the ICIP rights "that Indigenous people have, and want to have, to protect their traditional acts and culture" (Janke & Cubillo, 2019, p. 1). As discussed, inter-observer agreement was sought. The accuracy of the re-storied and transcribed dialogue was sought from Utopia Eastern Anmatyerr and Alyawarr participants, who emphasised the necessity to act the "right way" while on country.

These themes were later reviewed, refined, and named (Braun & Clarke, 2006) as three interconnecting themes: people, place, and time (see Figure 23). Of the nine initial themes, vague terms such as *bush* or *ones* were excluded as, in context, *bush* referred to animals and plants as bush animals or bush plants; similarly, *ones* in the transcript referred to *green ones* (unripe akatyerr) or *yellow ones* (ripe akatyerr). *Brown ones* referred to dried akatyerr, which was not considered a theme as it was the topic of investigation. *Turkey* was used in reference to bush animals and was only used in one file. *People, places*, and *time*, however, were accepted as the broad themes because these related directly to how the akatyerr harvest knowledge shared was to be treated. For instance, the transcription suggested how people should act while on country and in relation to kinship. The analysis revealed how the contextualisation of measurement learning informed by Indigenist research guidelines enacted "ceremony" by bringing people together to "occupy the same space" (Wilson, 2008, p. 87). For Māori, enacting ceremony "eliminates the space between people" (Wilson, 2008, p. 87). When "coming together after a struggle" (or "Makarrata"; Referendum Council, 2017, para. 10), school support staff, Utopia Eastern Anmatyerr and Alyawarr community members, assistant teachers, students, and I enacted a ceremony. Coming together to participate in ways to counter colonialism (Rigney, 1999), we, in a sense, enacted ceremony – in that we occupied the same space. Conducting research while respecting the cultural obligations of the adult participants ensured the relations among all the participants, and between the participants and the land at the akatyerr site, were maintained.

2.2 *The* Right Way *to Decolonise Measurement Education*

The *right way* to conduct the akatyerr study was revealed when considering the interconnecting relationships between people, place, and time. The people represented the students, assistant teachers, community members, and me while conducting the research. The place was the land visited for two-way dialogue on akatyerr harvest to occur. The concept of time was that which was reflexively referred to within audiovisual recorded dialogue. Respecting the importance of the right configurations of people, place, and time (Christie et al., 2010), and responding to the aspirations and directions of the adult participants at the akatyerr site, created the space for the two-way dialogue and intergenerational transfer of knowledge to occur.

Further, within the shared space, acting in the right way enabled adult participants to include plant knowledge and other concepts interrelated to akatyerr in the students' measurement learning. This also permitted the "right that Indigenous people have, and want to have, to protect their traditional [plant knowledge]", that is, their ICIP (Janke & Cubillo, 2019, p. 1). Additionally, the shared space created during subsequent member-check meetings enabled adult participants to check the accuracy of the transcripts and redact any secret or sacred knowledge. The permission to access, or instruction to redact, knowledge transferred in yarning circles protected the right for adult participants to manage "any information [in the transcriptions] … that is told … for publication purposes" (Janke & Cubillo, 2019, p. 3). Figure 23 illustrates the interconnecting relationships between the three themes of people, place, and time that emerged from the analysis of the yarning circle dialogue and my field notes.

The broad themes – people, place, and time – are represented in Figure 23 as interconnected circles. The interlocking section in the centre of the figure represents how the dialogue on sustainable akatyerr harvest is to be understood.

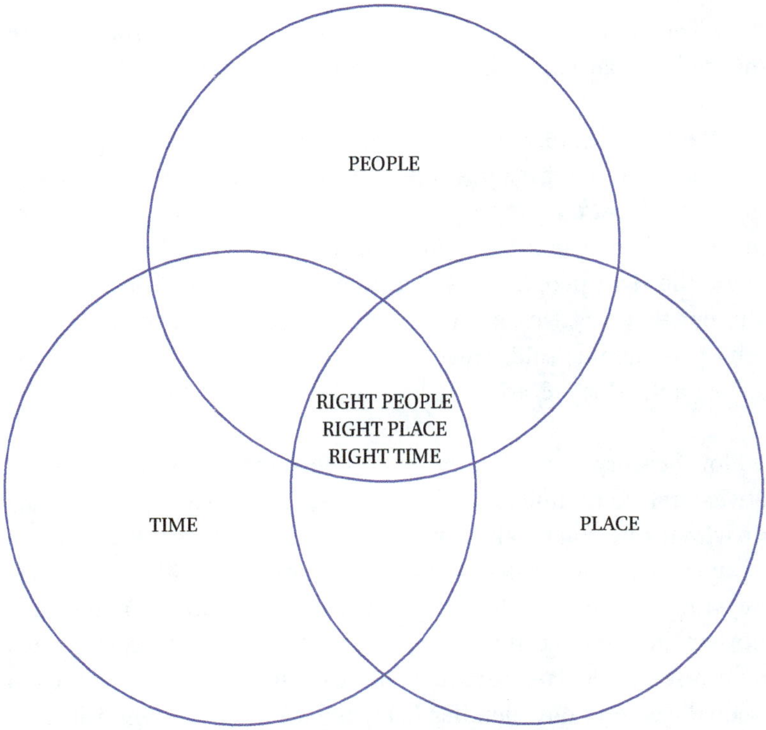

FIGURE 23 The right way to conduct two-way dialogue on akatyerr

As a non-Indigenous participant on land that I did not know, I was unable to contextualise measurement curriculum and pedagogic practice to place without collaborating with the right people. This was because I did not have the skills to listen to, recognise, and speak for the "place-based relations" – explicitly, the relations between plants, animals, and people (Harrison et al., 2019, p. 246) when at the akatyerr site. However, I was able to promote student agency over measurement learning by collaboratively acting in the shared space with the right people because I understood and viewed curriculum production as a social practice. I collaboratively acted with the right people: those who through kinship, experience, and skills could listen to and respond to the curriculum enacted when visiting the akatyerr site. Specifically, the right people had an ability to understand "place-based relations" (Harrison et al., 2019, p. 246) and speak for the knowledge concepts interrelated to akatyerr harvest when at the right place and with reference to the right time.

2.2.1 Right Place

Determining the right place to conduct the akatyerr harvest site visit occurred during yarning circles held regularly inside a study centre at a tertiary institute in community. The interaction I had with a community member and a parent of one of the students participating, is retold through the following field notes:

> There were three significant akatyerr growing areas. We discussed the kinship related to the three possible sites. As the most significant patch was considered, at *that* [emphasised] point, Lucy came over to the yarning circle. She had been sitting, listening, before that. I said to the parent, "Who are the right people to ask?" I prompted further, "That's Henry's country out that way, isn't it? We've done a bush trip out there before". And she [the parent] said, "Yeah, it's his country, and Lucy's". (Nicole Boyd, field notes, May 16, 2022)

Lucy waited for the story to come along. She listened to the yarning circle discussion between the community member and me. Since Lucy had an existing relationship with the person speaking and was familiar with my role in a teaching context, she was able to begin to form a new relationship (Wilson, 2008, p. 86) with me within a similar context: a bush trip. A new relationship began to develop between Lucy and me as we discussed her kinship to the akatyerr site. In a sense, the research enacted ceremony in this first yarning circle since the space between the community member, Lucy, and me began to diminish. In so doing, we strengthened our relationship. For Lucy, this new relationship began at the point where her name was mentioned in the yarning circle. Lucy offered advice about who were the right people to invite and made suggestions about how we could visit the chosen site. Once the right place was determined, Lucy and other people related to the site by kinship (such as Louisa and Sharmic) started to join in subsequent yarning circle discussions. Lucy, Louisa, Sharmic, each with relationships to this place, began to take control of how the bush trip to the chosen akatyerr site would be enacted. The ongoing consultation among the Utopia Eastern Anmatyerr and Alyawarr participants during two further yarning circles ensured the relationships among people, and between people and the land, were maintained (Wilson, 2008, pp. 86–87) the right way.

2.2.2 Right People

The right people – the adult participants such as Lucy, Louisa, and Sharmic – facilitated the right way to conduct the two-way dialogue on akatyerr. The right people appointed themselves as the appropriate cultural authority for the site chosen. They guided further decisions regarding appropriate people to also participate in yarning circles at the akatyerr site. The right people had experience and skills in perceiving the relations between the people, animals, and plants when at the akatyerr site. That is, the right people could listen to and recognise "place-based relations" (Harrison et al., 2019, p. 246) when at the akatyerr site. The Utopia Eastern Anmatyerr and Alyawarr adults, who have a right to practise traditional acts such as harvesting akatyerr, are able to perceive the right way of transferring knowledge concepts interrelated to this plant food. They have the cultural authority to instruct the people present at the site since their kinship to the site "encodes the relationships between people ... and their traditional country, the plants and animals that originate in that country" (Green et al., 2019, p. 422). For instance, Louisa instructed how each participant was to sit in the yarning circle in relation to the land while conducting two-way dialogue. This understanding of the relationship of each member of the yarning circle with respect to each other and the land was encoded through her kinship.

2.2.3 Right Time

As discussed above, the akatyerr harvest site we visited was considered the right place to participate in a bush trip. This was because the site visited was identified and overseen by the right people. In addition, when at the akatyerr site, the two-way dialogue enacting a transfer of knowledge concepts interrelated to akatyerr occurred with reference to the right time. That is, the adults' responses to the students' questions on their interests in knowledge concepts interrelated to akatyerr reflected a reference to the time that the questions were asked, the use of the site in the past, and a generalised rule for the present and ongoing time.

For Yolŋu people (north-east Arnhem Land), when the time is right, "the place tells the story when its people are there" (Christie et al., 2010, p. 77). Similarly, for Utopia Eastern Anmatyerr and Alyawarr, the use of the akatyerr site provided opportunities to understand akatyerr harvest according to the "appropriate configurations of people-place-moments" (Christie et al., 2010, p. 77). After Louisa arranged the adult and student participants to sit in a certain direction with respect to the land, she guided the students to ask their questions. During the yarning circle, a student participant asked Louisa how much it rains. When Louisa answered, she considered her response first with respect to her perception of the land at the time she was asked:

> Louisa: [Looks up into the sky] Sometimes we get rain. When clouds form, we see big rain. [Then generalising] In summertime, hot time, cold time you can get that [rain] – it can happen like that. [The present and ongoing time] Rain can fall at any time. (Yarning circle BB1, June 13, 2019)

Louisa's dialogue on rain referred to the concept of time reflexively. First, Louisa considered what she was asked by observing the sky at that moment –

there were no clouds. Then, Louisa generalised about rain during the seasons. Finally, from her experience and skills when listening to the land, she explained her perception that rain at the site can occur at "any time". The story of how it rains at the akatyerr site comes along while Louisa is sitting on her ancestors' land, where she and other forebears have harvested an important food source, the akatyerr. Similarly, Christie et al. (2010, p. 72) purport that "when their time is right ... the land is talking to them, [the knowledge] is ready to be told to the younger generation".

The student recording Louisa's responses was aware of the importance of the relationships she had among the adult participants who instructed her. The student's parent had consented to her participation in the research since the parent was also aware of the people, place, and time configurations that were arranged for the akatyerr site visit. This student, while recording the answers to her questions on akatyerr harvest knowledge concepts, was enacting "ceremony" (Wilson, 2008). That is, the student strengthened her relationship with the right people, who participated to share important cultural knowledge with her. The student listened attentively and respectfully to the responses. She demonstrated respect for the cultural authority of the speaker. While listening, this student reinforced her connection to her kinship with the adults and the land that she sat on. She reaffirmed the importance of her identity. The pedagogical practice of questioning elders on akatyerr knowledge concepts of interest to her supported her sense of self-efficacy. Strengthening student identity and sense of self-efficacy builds resilience. The care taken to provide culturally responsive curriculum supported the student to traverse between the cultural authority of her kin and other community members and the Western epistemologies embedded within the akatyerr investigation (quadrat survey design). The pedagogical practices of listening and responding to the right configurations of people, place, and time (Christie et al., 2010) within the yarning circle and quadrat survey supported student agency with measurement learning. Moreover, these pedagogical practices permitted students to access and maintain knowledge concepts interrelated to akatyerr, while constructing multiple representations of their measurement learning.

The right concept of time associated with the knowledge transfer was also dependent on who was being asked. This was because the adults' responses were dependent on their kinship relations to the site. Men and women have different responsibilities to manage the site. The responses to questions asked by the students reflected these gendered differences. For instance, in response to the students' question on whether to burn the akatyerr site, the answers reflected both the time the question was asked and who was answering. The male land manager participating in the yarning circle, Randall, considered the question of whether to burn or not, then gestured to the appropriate place to burn and answered, "Yeah" it is appropriate to burn "a little bit". However, Louisa, a female elder participating in a separate yarning circle, strongly disagreed with burning the site when asked, answering emphatically:

> Louisa: No!! ... Traditional owners and land managers [men] burn old grass to bring up green shoots to look after this country. Someone might accidentally burn off the wrong way – mistakenly. (Yarning circle bb1, June 13, 2019)

The right way to manage this akatyerr growing site could not be understood without considering the interconnections between the right people, place, and time. Without understanding the relationships of each participant to the site, and the responsibilities encoded by their kinship, it would be possible to act in the wrong way. Therefore, the concept of the right way is dependent on understanding, responding to, and maintaining the cultural obligations encoded through kinship. This is to ensure people, when at the akatyerr site, act the right way. The right way to act includes understanding the reference to time, both at the time the questions are asked and by listening to and considering the responses according to the right time reflexively referred to during the two-way dialogue exchanged.

2.3 *The Student Interests in Akatyerr Knowledge Concepts*

The inclusion of student voice in the study strengthened the relevance of curriculum connections made across learning areas as it related directly to their local context. The interconnectedness of Indigenous language and culture content and Aboriginal and Torres Strait Islander science elaborations adapted to sustainable wild akatyerr harvest as a local context supported the cross-curriculum contextualisation of akatyerr knowledge with measurement learning. In so doing, the curriculum choices strengthened connections between students' home language and culture and school language and culture as articulated in the Australian Curriculum (ACARA, 2020a, 2020b, 2020c; NTDOE, 2017).

2.3.1 Validating Student Interests

The investigation into akatyerr began by establishing which of the knowledge concepts were of interest to the students. The "written conversations" from student responses to their interests in akatyerr knowledge concepts were essential for giving voice to students (Bessarab & Ng'andu, 2010, p. 38). Students conveyed their akatyerr knowledge concept interests through their written and spoken responses. Written data from seven students and their assistant teacher (Sharmic) were collected. Four students chose not to participate, and one student found no concepts of interest to them. Each participant provided "information rich" (Woolley et al., 2013, p. 92) insight on interests relating to akatyerr cultural knowledge. The data from students' written dialogue were coded diversely (34 codes were assigned). That is, I was interested in dialogue about student interests in akatyerr, so I coded these interests without consideration to previous research (Braun & Clarke, 2006, p. 84). The knowledge concepts of interest were initially pooled per student; 13 patterns (themes) emerged, which are depicted in the right panel in Figure 24.

Figure 24 also depicts the frequency of concepts of interest to the students and one of the classroom assistant teachers, Sharmic. I listened to the written responses from the assistant teacher to guide which of the 13 themes could be used for the akatyerr investigation at the site. Six themes were selected initially. The interests *burning*, *rainfall*, *water places*, *land management*, *harvesting rules*, and *plant names*, emerged as the most frequently identified coded references. *Bush animals* also emerged through gathering data while participants were engaged in "different forms of conversation" (Bessarab & Ng'andu, 2010, p. 38). That is, how to promote bush animals was considered an important akatyerr knowledge interest which emerged during direct dialogue (conversation).

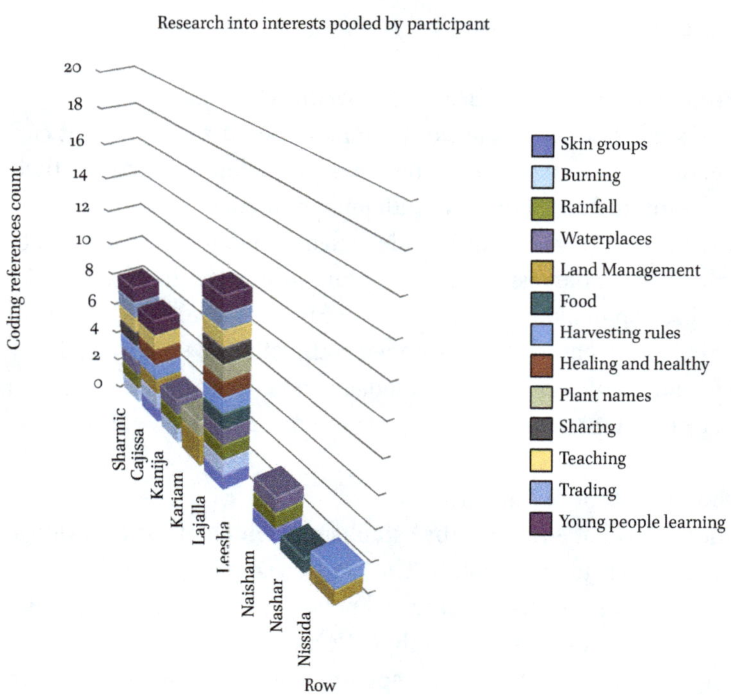

FIGURE 24 Student interests in akatyerr knowledge concepts pooled by each student participant and classroom assistant teacher, Sharmic

I listened to the responses from the assistant teacher to guide which of the 13 themes could be used for the Akatyerr investigation at the site. During the analysis of coded data, it was important at this early stage to give unequal attention to the interests identified from the assistant teacher. I considered all of the initially identified codes and data extracts that "demonstrated" each code from the data items provided by students and their classroom assistant teacher (Braun & Clarke, 2006, p. 89). Then, I critically reflected on the "tensions and inconsistencies" that emerged across and within their data items (Braun & Clarke, 2006, p. 89). I reflected which of the assigned themes that were part of the dominant story of the analysis could be used during the akatyerr site visit. For this, I gave unequal attention to the themes from the assistant teacher as they represented the cultural authority of an adult.

The cultural authority of an adult was considered superior to that of the students. This was essential to avoid making a mistake during the akatyerr site visit. The themes reflected in the coded data provided by Sharmic, the classroom assistant teacher, were valued above all themes generated from the student participants' data items. Where Sharmic's data were the same as the dominant story provided by the students, these interests were included. Where Sharmic's coded data differed from the dominant story provided by the students' coded data, these interests were not included. Accounts of the data that departed from the assistant teacher's story were retained (in data sets); however, further investigation into these interests did not occur at the akatyerr site visit (Braun & Clarke, 2006, p. 89). This was decided so I was able to protect the rights of the adult participants from discomfort or harm. That is, if students were inquiring about potentially sacred or secret knowledge relating to Akatyerr, inquiry into this knowledge could inadvertently cause harm or discomfort to adult participants, who need to maintain cultural protocols regarding intergenerational transfer of knowledge. Consideration of cultural protocols was essential to respect the cultural authority of adult participants and maintain dignity among all participants while the students' investigation with adult participants occurred. Ultimately, seven themes were selected to be used as yarning circle question prompts (see Figure 25).

2.3.2 Student-Led Inquiry – Links in Australian Curriculum

The student interests were posed as question prompts in a research instrument designed for the yarning circles at the akatyerr harvest site. The questions asked during the yarning circles are made explicit in Figure 25. This figure links each of the seven student interests, posed as questions, to the relevant learning areas within the Australian Curriculum (and sub-strands).

The students' interests were contextualised to the Australian Curriculum sub-strands of Country and Land, Natural Environment, People and Kinship,

Interests related to people	Australian Curriculum links
· **What are the akatyerr harvesting rules?**	Country and Land; People and Kinship; Measurement: mass, volume, rate
· **What are the plant names?**	Biology; Use of first language
Interests related to country	**Australian Curriculum links**
· **Where are the water places?**	Natural Environment; Earth Sciences
· **How much does it rain?**	Natural Environment; Measurement: capacity, area, volume
Interests in land management	**Australian Curriculum links**
· **Should we burn this country?**	Use and influence of science
· **How do we promote bush animals?**	Biology
· **How do we care for this country?**	Natural Environment; Measurement: perimeter, area, length

FIGURE 25 Student interests in akatyerr knowledge concepts related to people, country, and land management contextualised to the Australian Curriculum as question prompts for yarning circles

Measurement, Biology, Earth Sciences, Use of Indigenous language and culture, and Science Elaborations for Aboriginal and Torres Strait Islander cross-curriculum priorities. There were potentially also opportunities for students to learn measurement attributes while making connections across the mathematics sub-strands of number and place value (see Table 2). Additional opportunities for Year 8 curriculum links to volume of prisms and geometric reasoning, for instance, are provided in Chapter 8.

The next chapter elaborates how students' interests were adapted and trialled within Goompi measurement pedagogical cycles to explore measurement attributes (Matthews, 2006, 2020a, 2020b). The chapter describes how students experienced and explored the attributes, the relationship between attributes, and the big pedagogical ideas within each trial of the Goompi pedagogical cycle (Matthews, 2006, 2020a, 2020b).

References

Aboriginal Land Councils and Northern Territory Government. (2018). *The Barunga Agreement 8 June 2018: A memorandum of understanding to provide for the development of a framework for negotiating a treaty with the First Nations of the Northern Territory of Australia*. Retrieved December 9, 2022, from https://dcm.nt.gov.au/__data/assets/pdf_file/0003/514272/barunga-muo-treaty.pdf

Australian Curriculum, Assessment and Reporting Authority. (2020a). *Australian curriculum cross-curriculum priorities: Aboriginal and Torres Strait Islander histories and cultures*. Retrieved January 11, 2023, from https://australiancurriculum.edu.au/f-10-curriculum/cross-curriculum-priorities/aboriginal-and-torres-strait-islander-histories-and-cultures/

Australian Curriculum, Assessment and Reporting Authority. (2020b). *Australian curriculum: Learning continuum of numeracy: Using fractions, decimals, percentages, ratios and rates (Level 5)*. Retrieved January 11, 2023, from https://bit.ly/3GzjB4i

Australian Curriculum, Assessment and Reporting Authority. (2020c). *Australian curriculum: Mathematics (Version 8.4) – Year 8 level description*. Retrieved January 11, 2023, from https://bit.ly/3GB8zM7

Averill, R., Te Maro, P., Anderson, D., Easton, H., & Taiwhati, M. (2020). Bicultural mathematics teacher education and research. In C. Nicol, J. Archibald Q'um Q'um Xiiem, F. Glanfield, & A. J. (Sandy) Dawson (Eds.), *Living culturally responsive mathematics education with/in Indigenous communities* (pp. 113–135). Brill. https://doi.org/10.1163/9789004415768_006

Bessarab, D., & Ng'andu, B. (2010). Yarning about yarning as a legitimate method in Indigenous research. *International Journal of Critical Indigenous Studies, 3*(1), 37–50. https://doi.org/10.5204/ijcis.v3i1.57

Braun, V., & Clarke, V. (2006). Using thematic analysis in psychology. *Qualitative Research in Psychology, 3*(2), 77–101. https://doi.org/10.1191/1478088706qp063oa

Central Land Council. (2018, August). Pertame school comes to language's rescue. *Land Rights News Central Australia, 8*(2), 9. Retrieved December 9, 2022, from https://www.clc.org.au/land-rights-news/

Christie, M., Guyula, Y., Gotha, K., & Gurruwiwi, D. (2010). The ethics of teaching from country. *Australian Aboriginal Studies*, 2, 69–80. Retrieved January 11, 2023, from https://search.informit.org/doi/abs/10.3316/ielapa.767045875963420

Corntassel, J. (2012). Re-envisioning resurgence: Indigenous pathways to decolonization and sustainable self-determination. *Decolonization: Indigeneity, Education & Society, 1*(1), 86–101. Retrieved January 11, 2023, from https://jps.library.utoronto.ca/index.php/des/article/view/18627

Dobson, P. V., Walsh, F., & Sati, W. (2008). *Anpernirrentye: Relationships between bush foods, creation laws, people, country and all things, illustrated by three plants* [DVD]. Desert Knowledge Cooperative Research Centre.

Finlay, S. M. (2020, May 27). *Where do you fit? Tokenistic, ally – or accomplice?* Croakey Health Media. Retrieved December 9, 2022, from https://www.croakey.org/where-do-you-fit-tokenistic-ally-or-accomplice/

Finlay, S. M. (2021, January 26). *How to be a good Indigenous ally*. SBS NITV Politics. Retrieved December 9, 2022, from https://www.sbs.com.au/nitv/article/how-to-be-a-good-indigenous-ally/7czmek946

Green, J., Blackman, D., & Moore, D. (2019). *Alyawarr to English dictionary* (2nd ed.). IAD Press.

Harrison, N., Tennent, C., Vass, G., Guenther, J., Lowe, K., & Moodie, N. (2019). Curriculum and learning in Aboriginal and Torres Strait Islander education: A systematic review. *The Australian Educational Researcher*, *46*(2), 233–251. https://doi.org/10.1007/s13384-019-00311-9

Janke, T. (2021). *True tracks: Respecting Indigenous knowledge and culture.* NewSouth Publishing.

Janke, T., & Cubillo, C. (2019). *Towards an ICIP (Indigenous cultural and intellectual property) strategy for your work at Walyarra School* [pseudonym]. Terri Janke and Company.

Matthews, C. (2006). *Care for country initiative* [Unpublished manuscript]. Griffith University.

Matthews, C. (Guest Host). (2020a). Making a connection to culture (No. 38) [Podcast episode]. In *The random sample.* ARC Centre of Excellence for Mathematical and Statistical Frontiers (ACEMS). Retrieved December 15, 2020, from https://acems.org.au/podcast/episode-38-chris-matthews

Matthews, C. (2020b). *Teaching mathematics from a cultural perspective.* State of Victoria (Department of Education and Training). Retrieved December 19, 2022, from https://www.education.vic.gov.au/school/teachers/teachingresources/discipline/maths/Pages/research_culturalperspective.aspx

Merne Altyerre-ipenhe [Food from the Creation time] Reference Group, Douglas, J., & Walsh, F. (2011). *Aboriginal people, bush foods knowledge and products from central Australia: Ethical guidelines for commercial bush food research, industry and enterprises* (Desert Knowledge Cooperative Research Centre Report No. 71). Ninti One Limited. Retrieved January 11, 2023, from https://www.nintione.com.au/resource/NintiOneResearchReport_71_BushFoodGuidelines.pdf

Northern Territory Department of Education. (2017). *NT Indigenous languages and cultures: First Languages Maintenance (L1M).* Northern Territory Government. Retrieved January 11, 2023, from https://education.nt.gov.au/__data/assets/pdf_file/0020/715232/ntilc-l1m-nov18-03.pdf

Referendum Council. (2017). *Uluru statement from the heart.* Retrieved December 8, 2022, from https://ulurustatement.org/the-statement/

Rigney, L. I. (1999). Internationalization of an Indigenous anticolonial cultural critique of research methodologies: A guide to Indigenist research methodology and its principles. *Wicazo Sa Review*, *14*(2), 109–121. https://doi.org/10.2307/1409555

Scherer, P., & Steinbring, H. (2006) Noticing children's learning processes – Teachers jointly reflect on their own classroom interaction for improving mathematics. *Journal of Mathematics Teacher Education*, *9*, 157–185. https://doi.org/10.1007/s10857-006-0004-7

Schwandt, T. A., Lincoln, Y. S., & Guba, E. G. (2007). Judging interpretations: But is it rigorous? Trustworthiness and authenticity in naturalistic evaluation. *New Directions for Evaluation*, *2007*(114), 11–25. https://doi.org/10.1002/ev.223

Strauss, A. (1987). *Qualitative analysis for social scientists*. Cambridge University Press.

Turner, M. (2005). *Bush foods: Arrernte foods from Central Australia* (2nd ed.). IAD Press.

United Nations. (2007). *United Nations declaration on the rights of Indigenous peoples*. Australian Human Rights Commission. Retrieved January 11, 2023, from https://humanrights.gov.au/our-work/un-declaration-rights-indigenous-peoples-1

Wilson, S. (2008). *Research is ceremony: Indigenous research methods*. Fernwood Publishing.

Woolley, T., Sivamalai, S., Ross, S., Duffy, G., & Miller, A. (2013). Indigenous perspectives on the desired attributes of medical graduates practising in remote communities: A Northwest Queensland pilot study. *The Australian Journal of Rural Health*, *21*(2), 90–96. https://doi.org/10.1111/ajr.12018

YuMi Deadly Centre. (2014). *YuMi Deadly Maths AIM overview: Objectives, acceleration framework, pedagogy, culture and school change, implementation options*. Queensland University of Technology. Retrieved December 9, 2022, from https://research.qut.edu.au/ydc/wp-content/uploads/sites/181/2019/12/AIM-Overview-2014.pdf

CHAPTER 5

The Goompi Model

Minjerribah, the sand island in Moreton Bay (Queensland), is "a giant in the sun", as immortalised by Noonuccal poet and activist Oodgeroo (Noonuccal, 1977). Professor Chris Matthews is a mathematician of Quandamooka (Moreton Bay) heritage with strong connections to Minjerribah (Stradbroke Island) and the Minjerribah community (Matthews, 2009, p. 46). Chris first conceptualised the Goompi pedagogical model on Minjerribah (Matthews, 2020) – looking across its "belly brimming with waters so cool" (Noonuccal, 1977). Chris defined the Goompi model as iterative pedagogical cycles which incorporate the cultural bias of the students (Matthews, 2006, 2009). Chris explains:

> The model is called the Goompi Model because it is the traditional name of the Country [where] Dunwich State School resides. The name keeps the connection back to Country: its place of origin. It is also about keeping all the connections alive that are represented in the model to ensure that students experience mathematics that is part of their living world and part of who they are. (Matthews, 2020, p. 5)

This innovative pedagogy involves expressions of creativity and invokes a strong sense of Aboriginal identity (Matthews, 2009). I asked Chris if I could adapt the Goompi model as an intervention in my pedagogical practice. I chose to adapt the Goompi pedagogical cycle so that my teaching connected the culture of the learner to mathematics learning in pedagogically validating ways (Rigney, 2019).

1 Goompi Iterative Pedagogical Cycles in Four Phases

The pedagogical cycle was articulated through four phases: reality, abstraction, mathematics, and reflection (Matthews, 2006, 2009). In my study, each iterative pedagogical cycle began with the reality phase, which focused on the students' akatyerr harvest interests. In the abstraction phase, a range of creative acts, such as walking on country, helped to develop meaning relating to the notion of unit in measurement. In the mathematics phase, the notion of unit could be explored by focusing on a single attribute such as mass, capacity,

 | DOI:10.1163/9789004682757_005

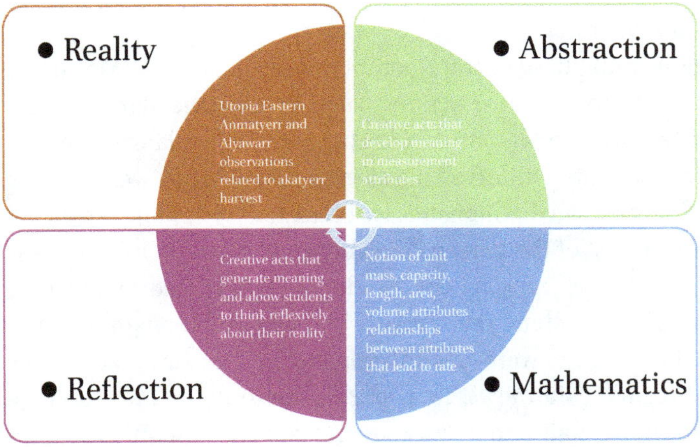

FIGURE 26 Phases of the Goompi pedagogical cycle (reality, abstraction, mathematics, reflection) adapted for the akatyerr harvest context from Matthews (2006)

length, area, or volume and/or the relationships between attributes. Students considered whether the mathematics reflected their observed reality. In the final phase, students reflected on the abstract forms of mathematics. Figure 26 depicts the four phases of the Goompi pedagogical cycle adapted for the akatyerr study.

1.1 *Reality – Perceived in Observations*

The Goompi pedagogical cycle began with "observations in a perceived reality" (Matthews, 2009, p. 47). This related students' mathematics learning to their interests and lived experiences. In this study, the students' real-life experience with akatyerr harvest was considered the "particular part" of their reality from which representations of mathematics were created (Matthews, 2009, p. 47). The quadrat survey method, for instance, allowed students to observe the akatyerr-growing land as discrete and measurable. The role of unit in turning a continuous or big place into the discrete and measurable was the big pedagogical idea explored (YuMi Deadly Centre, 2014, 2016). The "languages, culture and place of learning" for the akatyerr investigation provided the real-life context through which measurement learning could occur (NTDOE, 2017b, p. 5). What we perceive of our reality – in this case, for instance, the landscape, the distribution of the akatyerr plant, and the fruit yield – is dependent on our cultural bias. The Goompi pedagogical cycle takes account of the cultural bias. Matthews argues that our understanding of the "part of our reality" under investigation is socially constructed (Matthews, 2020, p. 5). From this perspective, the abstraction processes can be understood.

1.2 *Abstraction Processes*

Abstraction is the second phase of the Goompi pedagogical cycle (Matthews, 2006, 2020). Different forms of abstractions (such as language, dance, movement, drawing) offer different ways to express how students relate to the reality. In my study, the students, while walking along a transect line, were relating to the attribute of length. The mathematics is abstracted "to explore particular attributes and behaviours" of the perceived reality and "to communicate these ideas to others" (Matthews, 2009, p. 47). Students experienced length first, then recorded the distance along the transect line where the quadrat survey was conducted. The students were free to express "their cultural bias in the way they perceive reality" (Matthews, 2009, p. 48). The students' interests considered suitable to contextualise to measurement learning (land management activities such as monitoring invasive grasses) were explored.

1.3 *Mathematics Constructed*

The abstracted form of mathematics was interrogated and refined to ensure the "mathematical representation ... fits with the observed reality" (Matthews, 2009, p. 47) in this phase of the Goompi pedagogical cycle. The mathematics constructed by the students depicted the ways students explored and experienced the measurement attributes investigated and the associated big pedagogical ideas. The mathematics involved interacting with numeracy while estimating, factorising, and recording the measurement attributes such as the area of the quadrat covered in vegetation. The mathematics constructed captured and reflected the concerns related to the bush food industry, as Rayleen Brown argued:

> We need to concentrate on how do we support these women out on communities, how do we protect their intellectual property rights, or Indigenous ecological knowledge (IEK) as it's now known? How do they get a fair piece of the pie? (Brown, as cited in Finnane, 2018, para. 10)

To develop new measurement knowledge on what was considered a fair price – the amount of money traded for an akatyerr yield harvested – was solved as the dollars per kilogram ($/kg). A 100-gram sample of dried akatyerr harvest packaged by students (see Figure 27) was used in the classroom as a "concrete referent" to investigate proportional reasoning (Payne & Rathmell, 1977, p. 154). This example akatyerr product was integrated into the Big Business unit for the Indigenous Education Strategy Employment Pathways curriculum (NTDOE, 2016), as discussed in Chapter 3, Section 1. The akatyerr harvest depicted in Figure 27 is suitable to engage middle years Indigenous learnings since it

FIGURE 27
Dried akatyerr – a concrete referent packaged by students. The whole dried akatyerr pictured was harvested by students and staff
PHOTOGRAPH BY NICOLE BOYD

links Indigenous cultural knowledge and experiences in managing this valuable food source with measurement learning opportunities – simultaneously "working with and valuing" student difference (Lingard, 2005, p. 175).

1.4 *Reflection – on the Ways to Experience Attributes*

The critical reflection phase of the Goompi pedagogical cycle is "underpinned by the cultural bias within the abstraction process and the observer's perception of reality" (Matthews, 2009, p. 48). Reflection on the abstract form of mathematics created involves recognising that the meaning "attached" to a "symbol or group of symbols" reflects the cultural bias of the observer (Matthews, 2009, p. 48). Students engaged with the mathematisation of the cultural practice of harvesting akatyerr by exploring measures of length and area at the akatyerr harvest site. Students reflected on visual, oral, and numeric text to estimate percentage of an object and to solve area of an authentic context.

Being flexible and *generalising* are two examples of reflexive strategies students could use to check if their mathematics characterised the reality they observed. For instance, by being flexible or experiencing the attributes in many ways, students could determine if how they walked, drew, photographed, symbolised, and spoke about attributes was the same as what they observed. Students could reflect on the creative acts chosen to autonomously and flexibly represent the patterns observed in the quadrat frame. Generalising occurs when students represent the attribute of area, for example, by creating a symbol pattern (o) in a spreadsheet to reflect the measured percentage of 1 m^2 covered by akatyerr at each quadrat point (see Figure 33; see also Chapter 7, Figure 42). The Goompi pedagogical cycle supports students to "actively construct their own knowledge" (Sunzuma & Maharaj, 2019, p. 13) as articulated in subsequent iterations of this pedagogic cycle.

2 Goompi Pedagogical Cycles: Akatyerr Quadrat Study Iterations

The Goompi measurement pedagogical cycle (Matthews, 2006, 2020) was trialled during the akatyerr quadrat study. The trial implemented the curriculum content described in Table 2. As discussed in the previous section, each iteration of the Goompi pedagogical cycle began with the reality of the Utopia Eastern Anmatyerr and Alyawarr community: the observations related to akatyerr harvest. The assistant teacher, Sharmic, had revealed in yarning circles the capacity of dried akatyerr harvest collected (in buckets) from 100 cm × 100 cm areas at the site in the past. She estimated that for each square metre students may harvest "one bucket or two bucket or even three" (Sharmic) of dried akatyerr. This informed part of the reality under investigation at the akatyerr site. However, at the time of the investigation no fruit were available. The abstraction involved creative acts such as pointing towards the direction a transect line would be located (see Figure 20). Walking and laying the 50-metre transect line enabled students to experience length partitioned into 1-metre intervals. The mathematics involved acting with numeracy and estimating, factorising, and recording the measurement attributes for length and area. The reflection involved using reflexive strategies through creative acts such as photographing, drawing, and discussion of proportional reasoning applied to the percentage of an object (vegetation cover) to solve area of an authentic context. Real-world representations, such as photographs and audiovisual recordings, as well as diagrams and drawings, assisted students to move gradually from the familiar or real world to the unfamiliar or abstracted Western mathematics in iterative Goompi pedagogical cycles (see Chapter 7, Figure 42; Matthews, 2006, 2020; YuMi Deadly Centre, 2014).

The akatyerr cultural knowledge concepts – for instance, harvesting protocols or rainfall (see Figure 12) – provided "the big ideas" for students, assistant teachers, and teachers "to work with" (NTBOS, 2017, p. 28).

Student interests together with the voice of Utopia Eastern Anmatyerr and Alyawarr community determined that akatyerr harvest was a suitable real-life context for measurement learning. The students' interests were embedded in the design of the akatyerr study to promote student access to and agency over their measurement learning. The students' interests informed the changes to pedagogic practices to reflect their voice in their education. I adapted and trialled a Goompi measurement pedagogical cycle (Matthews, 2006, 2020) to embed students' interests such as caring for country. Students investigated changes to plant density across the akatyerr-growing landscape by estimating the percentage cover of akatyerr vegetation and other plants growing at the site. The ways the Goompi pedagogical cycle and its four phases – reality, abstraction, mathematics, and reflection – were adapted to the akatyerr context were described broadly in Section 1 and shown in Figure 26 (Matthews, 2006, 2020). The study identified that area and length could be adapted to this pedagogical cycle. The images and text in the following subsections depict each phase of the cycle conducted at the akatyerr site in more detail.

2.1 *Reality*

The Goompi cycle trialled in the study began with the students' interests and lived experiences with akatyerr harvest. The students' real-life experiences with akatyerr were considered the reality. The "languages, culture and place of learning" for the akatyerr investigation provided the real-life context through which the measurement learning could occur (NTDOE, 2017b, p. 5). The cultural authority for the Indigenous language and culture curriculum was the "communicative situation" of the Utopia Eastern Anmatyerr and Alyawarr Community (ACARA, 2020). The yarning circle processes that supported the contextualisation of akatyerr knowledge with measurement attributes and associated big pedagogical ideas are depicted in Figure 28. The photograph in Figure 28 was taken during the first phase of this Goompi measurement pedagogical cycle, the reality.

2.2 *Abstraction*

During the abstraction phase the students' interests in caring for country were explored creatively. Students made decisions regarding how they would experience the measurement attributes. They began with a randomised trial to assign the quadrat points using a random number table. The image in Figure 29 depicts the students' agency when locating their randomly chosen quadrat points along the transect line.

The abstraction processes highlight the ways students related to the reality. The students pictured in Figure 29 walking along a transect line were relating to the attribute of length. The distance they walked to place the quadrat frame was a way the length measured was understood by the students. The students

FIGURE 28 Yarning circle at akatyerr harvest site: Students yarn with adults during the reality phase of the Goompi measurement pedagogical cycle
PHOTOGRAPH BY TIM WARE

FIGURE 29 Example creative act: walking the transect line to locate the quadrat point
PHOTOGRAPH BY TIM WARE

constructed mathematical meaning of the distance travelled along the transect line or length in metres (m) while they were walking. Students engaged with the big pedagogical idea of changing the notion of the akatyerr-growing landscape from continuous into discrete units of measurement such as length

(m) and area (m^2) while walking across the akatyerr site. Students engaged with this notion while they carried the quadrat frame along the 50 m transect line and then placed the quadrat frame at each quadrat point.

2.3 *Mathematics*

The mathematics constructed depicts the ways the students explored and experienced the measurement attributes of length and area and associated big pedagogical ideas. The ways students represented their understanding of akatyerr provided an opportunity for a broader discussion on techniques for eliciting student voice in teaching and learning measurement. Field methods that combine student participants' visual texts (like photographs and drawings) and dialogue (Cappello, 2005; Mayes, 2016; White et al., 2010) have enabled students to elicit data on their "own terms" (White et al., 2010). For instance, students experienced length first by choosing where to walk along the transect line to reach the distance where the survey was conducted (at random points 37 m, 13 m, 50 m, 27 m, 45 m, and 9 m along the transect line). This technique allowed students to explore the notion of unit on their terms – that is, the notion that units such as metres can break a continuous length into discrete parts, which can be found by walking along a 50 m measuring tape (transect line) and counting to locate the quadrat point.

Students also experienced area, and how it is related to division, on their own terms. Student pairs used a 1 m^2 quadrat frame, partitioned as a 5-cell × 5-cell array, as an overlay to assess the vegetation cover. Students engaged with division by "dividing the surface" of the akatyerr-growing landscape into equal parts (YuMi Deadly Centre, 2016, p. 69), in this case, 25 equal parts. That is, each cell (or subplot) divided the continuous square metre of landscape into 1/25 (or 4%). Students could count each subplot and multiply by 4, or count each row and multiply by 20, to estimate the percentage area covered in vegetation through the quadrat frame. For instance, for the *Typhonium* plant pictured in Figure 34, the example student artefact in Figure 30 reflects an estimation of vegetation cover based on 3 rows of the quadrat frame (multiplied by 20) or 15 subplots (multiplied by 4) covering the vegetation. That is, students could factorise basic arithmetic to estimate that the percentage of the one square metre of landscape in Figure 34 covered in *Typhonium* vegetation was 60%.

Using the same quadrat frame to conduct the investigation also reinforced another key measurement concept: "units must be the same size" when comparing measures (YuMi Deadly Centre, 2016, p. 69). Students could then make comparisons of whether the landscape surface had a larger or smaller percentage of square metres covered in vegetation and represent this through drawings on a grid, photographs, numeric symbols, and dialogue. For instance,

Quadrat number	Quadrat point	% Cover
1	1	0
2	8	0
3	39	60
4	4	0
5	2	0
6	7	0

FIGURE 30 Example student artefact recording two attributes: length (Quadrat point column) and area (% Cover column)
SOURCE: WORKSHEET COMPLETED BY STUDENT 2

students could compare the surface covered in grass, sand, stones, a rare and threatened plant (*Typhonium*), and akatyerr.

Mathematics can be represented in "several different but equivalent ways" using a range of creative tools (Davis, 2020, p. 153). Roger Antonsen, for instance, used a laser cut to demonstrate a mathematical concept using wood, paper, and mirrors, and argued, "in order to understand something we should look at it from different perspectives" (Davis, 2020, p. 153). In my study, students used photographs, drawings, symbols, and numeric and oral representations to understand the area of a quadrat covered in akatyerr and other vegetation and associated big pedagogical ideas. The study adapted the Goompi pedagogical cycle (Matthews, 2006, 2020) to enable the unique context of the learner to be the starting point for measurement learning experiences. The learning occurred in one visit to an akatyerr harvest place.

2.3.1 Akatyerr

Students readily participated in the measurement learning contextualised from their interests in, and experiences with, a familiar real-life context. For instance, students investigated environmental threats using the yarning circle question prompts and quadrat survey instruments developed for the akatyerr investigation. During the quadrat survey, students selected the unit of area (m^2) to investigate and recorded their observations of threats, such as invasive grass species, to the akatyerr food webs.

The curriculum adjustments enabled students to model real-world situations suited to Year 8 academic challenge. For instance, students were supported to move from a concrete image of akatyerr vegetation observed within the frame at quadrat point 49 to a representational image photographed (Figure 31), an image drawn on square grid paper, and numeric symbols.

The quadrat frame depicted in the photograph (Figure 31), shows there is no or 0% akatyerr vegetation within the frame. Since there was no vegetation cover growing at quadrat point 49, the students recorded this on grid paper (partitioned into 25 parts) as no or 0% vegetation (see Figure 32).

FIGURE 31 Student representation of area of the 1 m² quadrat covered in akatyerr vegetation (also mass and capacity) in photographed image of quadrat point 49 (visual representation)
PHOTOGRAPH BY STUDENT 4 [WORKING IN A PAIR WITH STUDENT 3]

FIGURE 32 Student representation of area of the 1 m² quadrat covered in akatyerr vegetation at quadrat point 49: no vegetation drawn on the graph paper (visual representation) and graph labelled 0% (numeric representation)
SOURCE: DRAWING AND NUMERIC REPRESENTATION BY STUDENT 3 [WORKING IN A PAIR WITH STUDENT 4]

Student 3's observation that 0% of the quadrat was covering akatyerr was also revealed in direct dialogue, shown in the excerpt below:

Nicole: [teacher points to quadrat] Any akatyerr?
Student 3: [emphatically] No.

Nicole: [smiles acknowledging] So what percentage is it then?
Student 3: Zero percent [student records the result].

What was not discussed but could be inferred by the student representations (graph and photograph) was that there was also no capacity or mass of dried akatyerr at this quadrat point. Hence, the quadrat study provided opportunities for students to construct their own understanding of measurement knowledge relating to area (% of m^2 covered in akatyerr plant), mass (g/m^2 of dried akatyerr fruit), and capacity (handful/m^2 of dried akatyerr fruit), or length (m of a transect line) in multiple ways.

Figure 33 depicts Student 3's representation of the attribute of area or the percentage of the m^2 quadrat frame covering akatyerr vegetation as a symbolised pattern (0) in a table. This table also depicts the student's representation of the attribute of length or distance along the transect line to each quadrat point randomly assigned. The two-way dialogue exchanged between the student and me, and the visual image of the quadrat as a reference point for the conversation, elicited the student's oral representation of her measure of area as "zero percent". The conversation traversed between the home-based language and plant knowledge with the question prompt of "Any akatyerr?" and the school-based language and culture with her response of "zero percent". The ability for her verbal representations to be applied reflexively back to the reality (the quadrat frame) confirmed the accuracy of her representation.

2.3.2 Typhonium

The following images and text retell the story of how the students contextualised measurement learning using a remnant rainforest plant, *Typhonium*. As there were no akatyerr plants growing along the 50 m transect, other plants were also

Quadrat Study: Akatyerr

Transect line number: 3

Quadrat number	Quadrat point	% Cover
1	37	0
2	13	0
3	50	0
4	27	0
5	45	0
6	9	0

FIGURE 33 Example student artefact recording two attributes: length (Quadrat point column) and area (% Cover column) relating to transect line number 3
SOURCE: WORKSHEET COMPLETED BY STUDENT 3 [WORKING IN A PAIR WITH STUDENT 4]

used to demonstrate the different ways students represented their measurement learning. Students observed and recorded their observations at quadrat point 39 on the research instrument designed for the akatyerr investigation. Conducting the quadrat survey involved holding the quadrat frame above the quadrat point (to protect the rare plant), as pictured in Figure 34. The students estimated the percentage cover through flexibly shifting between different representations during the abstraction phase of the Goompi pedagogical cycle. For instance, students recorded what they observed through a photograph and reflected their estimate of vegetation cover as a drawing. As with the akatyerr results, the *Typhonium* representations depict how the students experienced the attribute of area.

Students, working in pairs, provided their own representations of their experience with observing the vegetation cover within the 1 m^2 quadrat frame (Figure 34). Experience with measurement attributes, on their own terms, encouraged "autonomous learning practice" (Scherer & Steinbring, 2006, p. 160). Once the method for estimating percentage cover was modelled, students could then self-manage their measurement learning. For instance, students experienced the research instruments used to assess vegetation cover as a group before working in pairs. Figure 35 depicts the grid reference used to support students to estimate the percentage of 1 m^2 covered in *Typhonium*. Although Figure 35 depicts a 4 × 5 grid to reference the percentage cover, the

FIGURE 34 Student representation of area of the 1 m^2 quadrat covered in *Typhonium* (or pretty flower) vegetation in photographed image of quadrat point 39
PHOTOGRAPH BY TIM WARE

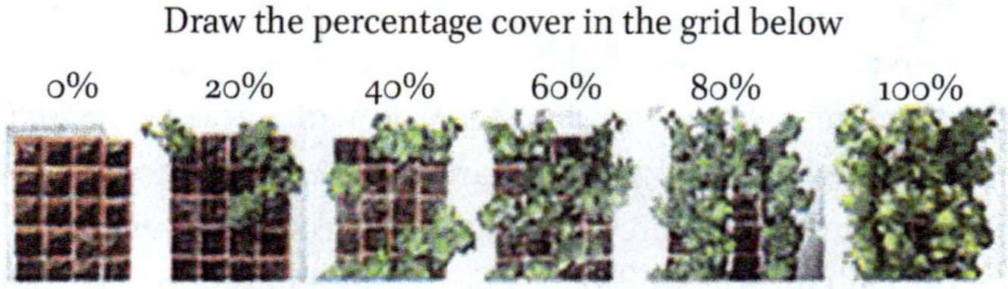

FIGURE 35 Grid reference for estimating percentage cover (adapted from Deakin University, School of Life and Environmental Sciences, n.d.)

illustration encouraged students to self-manage their learning and engage spatial visualisation skills.

The grid paper depicted in Figure 36 was used to represent the students' observations of vegetation cover inside the quadrat. The student observation of the number of cells or subplots within the quadrat covered in vegetation was represented as shaded squares on the grid paper. The representation on grid paper permitted students to apply their understanding of the distributive property of an array (e.g., as 3×5 or 5×3) and supported students to make the link between multiplication and area. The use of 5×5 cells in the quadrat and grid-paper partitions provided students with more flexible mental computation methods. It also demonstrated that the distributive property is not restricted to partitions involving tens. This aligned with the expectation for Year 8 mathematics students to be able to "apply the distributive law" and "factorise algebraic expressions" (ACARA, 2014a). Students were also given the opportunity to explore proportional reasoning.

The students reflected on the mathematics represented in their grid drawing (Figure 36) while holding the image of the grid reference (Figure 35) and

FIGURE 36 Example student artefact: Student 1's drawing of the area of the 1 m^2 quadrat covered in *Typhonium* vegetation at quadrat point 39. Student 1 worked in a pair with Student 2

the vegetation observed (Figure 34) in their mind, switching between representations. That is, students engaged their spatial visualisation skills to create and hold mental images of the vegetation "in space" then manipulate these images to reason about the proportion of vegetation cover observed through the quadrat frame (Battista et al., 2018, p. 197). Similarly, students held the image of the symbolised patterns in the worksheet (Figures 30 and 33), visual representations (photographs, grid reference, and drawings), and dialogue to see if what they observed accurately represented the reality (Battista, 2007; Battista et al., 2018). Visual representations, such as diagrams, helped students to build measurement conceptual knowledge.

Students observing vegetation cover were viewing the "spatial interval" within a subplot (Congdon et al., 2018, p. 33). Students engaged spatial reasoning skills, or their ability to recognise "spatial images" (grid, photograph, drawing) and "spatial objects" (vegetation, quadrat, area of the region in green, or subplot), and reflected on the relationships among these (Battista et al., 2018, p. 196; see also Battista, 2007). For instance, the percentage cover of *Typhonium* vegetation observed at quadrat point 39 (spatial object) was constructed as a diagram according to the perspective or spatial image held by the observer. Student 1 experienced the *Typhonium* vegetation at quadrat point 39 and recorded their estimation by shading 9 squares in the grid depicted in Figure 36. As stated earlier, each subplot represents 4% cover, so multiplying each square by 4 and rounding to the nearest estimate in the grid reference (Figure 35) would represent a 40% cover. The numeric value accurately reflects the photographed image and an estimate of 40% using basic arithmetic.

The experience with the attribute of area was different for Student 2. The student's estimation of the percentage of 1 m^2 covered in *Typhonium* vegetation, drawn on a grid, is depicted in Figure 37. The estimation drawn by Student 2 demonstrates a larger proportion of plant cover. She shaded 12 squares of the grid and symbolised this in the worksheet as 60%. The representation of 60% in the grid reference (Figure 35), with 12 subplots covered in vegetation, is similar to that which was drawn by the student in her own grid.

Both Student 1 and Student 2 constructed representations that critically reflected the new measurement learning they each observed, autonomously. Each student created their own representations with respect to their own proportional reasoning of the observed plant and images held from other acts of abstraction (referencing the grid guide, drawing, photographed image, and/or factorisation of basic arithmetic).

2.3.3 Grasses

Grasses were also assessed using the instrument designed for the akatyerr investigation. The student–teacher interaction on percentage cover of grasses

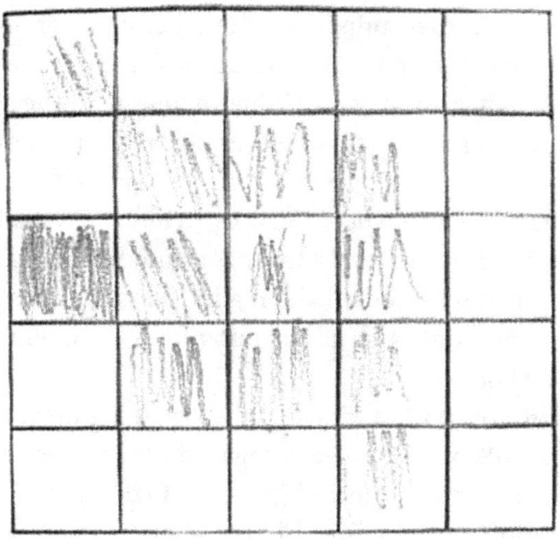

FIGURE 37
Example student artefact: Student 2's drawing of the area of the 1 m^2 quadrat covered in *Typhonium* vegetation at quadrat point 39

in the quadrat was based on the image in Figure 38. The figure depicts the student representation of the area of 1 m^2 quadrat covered in grass vegetation through a photographed image. The interaction the teacher and assistant teacher had with the students to elicit student voice was in response to the quadrat observed in Figure 38. The teacher also pointed to the grid reference (Figure 35) to guide the students' estimation of the percentage of 1 m^2 covered in grass vegetation. The opportunity to observe vegetation in a quadrat, through a photograph, and as a grid reference allowed students to construct new measurement learning on their own terms. Students, through these acts of abstraction, could generalise about the area of the quadrat covered in grass. The following narrates the interaction the teacher had with each student to elicit their individual understanding of the area covered in grass:

Nicole:	[holding the grid reference image up to students] What do you think it is? Is it 20, 40, 60, 80, or 100%?
Student 3:	60
Student 4:	40
Student 2:	[observed the discussion and chose not to contribute an answer]
Student 1:	[considered her answer] 60 [then reconsidered] 80

This subsection has demonstrated the ways students have come to represent their mathematics knowledge and skills in relation to the akatyerr knowledge concepts and measurement attributes such as length and area.

FIGURE 38 Example student artefact of a creative act: photographing the area of the 1 m^2 quadrat covered in grass vegetation at quadrat point 37
PHOTOGRAPH BY STUDENT 4 [WORKING IN A PAIR WITH STUDENT 3]

2.4 *Reflection*

Students engaged with the mathematisation of akatyerr knowledge by exploring measures of length and area. *Being flexible* and *generalising* are two examples of reflexive strategies students used to check if their mathematics characterised the reality they observed. For instance, by being flexible or experiencing the attributes in many ways, students could determine if how they walked, drew, photographed, symbolised, and spoke about attributes was the same as what they observed. As stated previously, students experienced length first by walking the transect line, then by recording the distance along the transect line where the survey was conducted or the quadrat point (37 m, 13 m, 50 m, 27 m, 45 m, and 9 m).

The students' understanding of their experience with measures of akatyerr (and other plants) reflected their *cultural bias*. For instance, a verbal representation of a student's observation at quadrat point 49 was that there was "no" akatyerr. In a worksheet, the student recorded this observation as the numeric symbol for "zero". The measured percentage of 1 m^2 covered by akatyerr at each quadrat point investigated by the student thereafter was also recorded numerically as zero (see Figure 33). The student generalised that, for the attribute of area, there was 0% of akatyerr per square metre. However, an adult participant found an akatyerr plant growing near this transect. The generalisation made by the student was an accurate reflection of her experience of the reality she observed. The reality is also reflected in the photograph in Figure 39, a semi-permanent soak (freshwater discharge) at the akatyerr site.

The reflection processes enabled the students' measurement meaning constructed through their akatyerr investigation (in movement, visual, numeric, and oral representations) to be applied back to the reality. Students critically reflected on the mathematics that they saw. The reflection on their reality involved flexibly moving between the representations of mathematics and what they saw in each quadrat frame. The Goompi pedagogical cycles used in my study, together with the students' previous lived experiences, enabled students to deepen their understanding of ways to care for country.

Another important consideration raised by the bush food harvesters was knowing how to obtain a fair price for akatyerr products. Each experience with adapted pedagogy to investigate akatyerr harvest helped to address this concern. Table 3 summarises each experience with measurement attributes adapted as Goompi pedagogical cycle trials (Matthews, 2006, 2020). The table

FIGURE 39 Semi-permanent soak (freshwater discharge) at the akatyerr site: a place to reflect the reality
PHOTOGRAPH BY TIM WARE

TABLE 3 Summary of each trial of the Goompi pedagogical cycle adapted for contextualising measurement attributes to akatyerr harvest

Reality	Abstraction	Mathematics	Reflection
Akatyerr harvest place visited in this study	Walking the transect	Length (50 m)	Being flexible
	Finding the quadrat point on the transect line	Length (m)	Being flexible
	Photographing quadrat	Area (% of m²)	Being flexible
	Referencing quadrat	Area (% of m²) Distribution of plants in grid guide (4 × 5 cells)	Being flexible
	Drawing quadrat	Area (% of 1 m²) Proportional reasoning of plant cover represented by students in grid (5 × 5 cells)	Being flexible
	Discussing % cover of other plants (*Typhonium*, grasses)	Area (% of m²) Estimating % cover of each cell – multiplying number of cells covered in vegetation by 4 Counting rows covered by plant and multiplying by 20 for each row	Being flexible Reversing
	Recording symbol patterns in spreadsheets	Area (% of m²) for each quadrat point	Being flexible Generalising
Akatyerr harvest place visited at a different time	Harvesting fruit	Volume (m³)	Being flexible Generalising
Akatyerr harvest at a different place at the same time	Holding akatyerr fruit in hand	Capacity (handful)	Being flexible Changing parameters
Akatyerr harvest at a different place at a different time	Heft of akatyerr fruit in collection bag	Mass (g, kg)	Being flexible Generalising
	Money traded	Rate ($/kg)	Generalising
	Yield harvested	Rate (g/m²)	Generalising
	Amount of time taken to collect 1 kg dried akatyerr	Rate (kg/hour)	Changing parameters Generalising

SOURCE: GOOMPI CYCLE ADAPTED FROM MATTHEWS (2006)

summarises the reality, abstraction processes, mathematics, and reflection or reflexive strategies involved for each measurement attribute contextualised to akatyerr harvest at the time of this study and at a different time. It provides details regarding the attributes – length, area, volume, capacity, mass – and relationships between attributes that lead to rate. For instance, measures of yield harvested (g/m^2) and money traded ($/kg) from previous lived experiences helped to develop new measurement knowledge about what is considered a fair price. As noted in Chapter 1 (Section 1.2), the price per bucketful based on lived experiences articulated by Ruby Morton and Lucky Morton was understood as $150/bucket (Finnane, 2018). However, this price was considered unfair considering the amount of time taken to collect dried akatyerr (3 or 4 days to collect 2 bucketsful). The price for whole dried akatyerr harvested can be traded for $35/kg to $45/kg depending on the market value (R. Brown, personal communication, June 2017).

The students' akatyerr investigation deepened their understanding of this important plant food and the landscape that supports its growth. The use of the Goompi pedagogic model provided "opportunities for students to engage" with sustainable akatyerr harvest and concepts of measurement within the conceptual framework of a "specific language and community" (NTBOS, 2017, p. 28), that is, according to the cultural bias of the student (Matthews, 2006, 2020). The akatyerr knowledge concepts were "not topics for students to learn about"; rather, akatyerr harvest provided "opportunities for students to learn through" (NTBOS, 2017, p. 28). The akatyerr knowledge concepts considered aligned with the Indigenous Languages and Cultures learning area of the Northern Territory curriculum framework (NTDOE, 2017a, 2017b). The content selected was based on "Aboriginal organisation of knowledge" such as in the language and cultural content areas: Country and Land, People and Kinship, and the Natural Environment (Douglas, 2011, p. 6; see also NTBOS, 2017, p. 28). New measurement learning was constructed through a respect for and responsibility to the cultural protocols or "rules and proper ways to behave on and treat Country" (NTDOE, 2017b, p. 17).

The study has reflected the need for "collaborative efforts of local and non-local staff" to transform pedagogic practices in remote education settings (Guenther, Disbray, & Osborne, 2015, p. 204). The students and local and non-local staff, together, created measurement curriculum contextualised to students' akatyerr interests. The study transformed pedagogical approaches in order to "build on and respect local languages and cultures" through "a sustained" research effort (Guenther, Disbray, & Osborne, 2015, p. 204). The following chapter examines the underlying narrative guiding this intervention in my pedagogical practices.

References

Australian Curriculum, Assessment and Reporting Authority. (2020). *Australian curriculum: Mathematics (version 8.4) – Year 8 level description.* Retrieved January 11, 2023, from https://bit.ly/3GB8zM7

Battista, M. T. (2007). The development of geometric and spatial thinking. In F. Lester (Ed.), *Second handbook of research on mathematics teaching and learning* (pp. 843–908). National Council of Teachers of Mathematics.

Battista, M. T., Frazee, L. M., & Winer, M. L. (2018). Analyzing the relation between spatial and geometric reasoning for elementary and middle school students. In K. S. Mix & M. T. Battista (Eds.), *Visualizing mathematics: The role of spatial reasoning in mathematical thought* (pp. 195–228). Springer. https://doi.org/10.1007/978-3-319-98767-5

Cappello, M. (2005). Photo interviews: Eliciting data through conversations with children. *Field Methods, 17*(2), 170–182. https://doi.org/10.1177/1525822X05274553

Congdon, E. L., Vasilyeva, M., Mix, K. S., & Levine, S. C. (2018). From intuitive spatial measurement to understanding of units. In K. S. Mix & M. T. Battista (Eds.), *Visualizing mathematics: The role of spatial reasoning in mathematical thought* (pp. 25–46). Springer. https://doi.org/10.1007/978-3-319-98767-5

Davis, D. (Ed.). (2020). *Illustrating mathematics.* American Mathematical Society. Retrieved December 9, 2022, from https://dianadavis.github.io/im/book.pdf

Deakin University, School of Life and Environmental Sciences. (n.d.). *SLE 101, practical manual for environmental techniques and monitoring.*

Douglas, J. (2011). *Learning from country: The value of country visits in remote schools for community engagement and natural and cultural resource management* (Desert Knowledge Cooperative Research Centre Report No. 69). Ninti One Limited. Retrieved January 11, 2023, from https://www.nintione.com.au/resource/NintiOneResearchReport_69_Thevalueofcountryvisitsinremoteschools.pdf

Finnane, K. (2018, March 2). Bush foods: How can wild harvesters get a piece of the pie? *Alice Springs News, 25*(2). Retrieved January 11, 2023, from https://alicespringsnews.com.au/2018/03/02/bush-foods-how-can-wild-harvesters-get-a-piece-of-the-pie/

Guenther, J., Disbray, S., & Osborne, S. (2015). Building on "red dirt" perspectives: What counts as important for remote education? *The Australian Journal of Indigenous Education, 44*(2), 194–206. https://doi.org/10.1017/jie.2015.20

Lingard, B. (2005). Socially just pedagogies in changing times. *International Studies in Sociology of Education, 15*(2), 165–186. https://doi.org/10.1080/09620210500200138

Matthews, C. (2006). *Care for Country initiative* [Unpublished manuscript]. Griffith University.

Matthews, C. (2009). Stories and symbols: Maths as storytelling. *Professional Voice, 6*(3), 45–50. Retrieved December 9, 2022, from https://www.aeuvic.asn.au/sites/default/files/PV_6-3.pdf

Matthews, C. (2020). *Teaching mathematics from a cultural perspective.* State of Victoria (Department of Education and Training). Retrieved December 19, 2022, from https://www.education.vic.gov.au/school/teachers/teachingresources/discipline/maths/Pages/research_culturalperspective.aspx

Mayes, E. (2016). *The lines of the voice: An ethnography of the ambivalent affects of student voice* [PhD thesis, University of Sydney]. Sydney Digital Theses. Retrieved December 9, 2022, from http://hdl.handle.net/2123/15274

Noonuccal, O. (1977). Minjerriba. *Meanjin Quarterly, 36*(4). Retrieved November 10, 2022, from https://meanjin.com.au/poetry/minjerriba/

Northern Territory Board of Studies. (2017). *Keeping Indigenous languages and culture strong: A plan for the teaching and learning of Indigenous languages and cultures in Northern Territory Schools.* Northern Territory Government. Retrieved January 11, 2023, from https://education.nt.gov.au/__data/assets/pdf_file/0012/413202/Policy_Keeping-Indigenous-Languages-and-Cultures-Strong-Document_web_updated.pdf

Northern Territory Department of Education. (2016). *Indigenous education strategy employment pathways: Country, culture, careers: Big business – Running an enterprise (version 3.1).* Northern Territory Government.

Northern Territory Department of Education. (2017a). *Guidelines for the implementation of Indigenous languages and cultures programs in schools.* Northern Territory Government. Retrieved January 11, 2023, from https://education.nt.gov.au/__data/assets/pdf_file/0004/471712/indigenous-languages-and-cultures-guidelines.pdf

Northern Territory Department of Education. (2017b). *NT Indigenous languages and cultures: First Languages Maintenance (L1M).* Northern Territory Government. Retrieved January 11, 2023, from https://education.nt.gov.au/__data/assets/pdf_file/0020/715232/ntilc-l1m-nov18-03.pdf

Payne, J., & Rathmell, E. (1977). Numeration. In J. Payne (Ed.), *Mathematics learning in early childhood: 1977 yearbook of the National Council of Teachers of Mathematics.* NCTM.

Rigney, L. I. (2019, September 20). A time of change and "super diversity" in early childhood education. *The Spoke: Early Childhood Australia's Blog.* Retrieved December 19, 2022, from https://thespoke.earlychildhoodaustralia.org.au/time-change-super-diversity-early-childhood-education/

Scherer, P., & Steinbring, H. (2006) Noticing children's learning processes – Teachers jointly reflect on their own classroom interaction for improving mathematics. *Journal of Mathematics Teacher Education, 9*, 157–185. https://doi.org/10.1007/s10857-006-0004-7

Sunzuma, G., & Maharaj, A. (2019). Teacher-related challenges affecting the integration of ethnomathematics approaches into the teaching of geometry. *EURASIA Journal of Mathematics, Science and Technology Education, 15*(9), Article em1744. https://doi.org/10.29333/ejmste/108457

White, A., Bushin, N., Carpena-Méndez, F., & Ní Laoire, C. (2010). Using visual methodologies to explore contemporary Irish childhoods. *Qualitative Research, 10*(2), 143–158. https://doi.org/10.1177/1468794109356735

YuMi Deadly Centre. (2014). *YuMi Deadly Maths AIM overview: Objectives, acceleration framework, pedagogy, culture and school change, implementation options.* Queensland University of Technology. Retrieved December 9, 2022, from https://research.qut.edu.au/ydc/wp-content/uploads/sites/181/2019/12/AIM-Overview-2014.pdf

YuMi Deadly Centre. (2016). *YuMi Deadly Maths: Measurement – Prep to Year 9.* Queensland University of Technology. Retrieved December 9, 2022, from https://research.qut.edu.au/ydc/wp-content/uploads/sites/181/2019/12/YDM-Book6-Measurement-2016.pdf

CHAPTER 6

Narrative: The Experience and Story of Decolonising Mathematics Education

Narrative inquiry as a methodology generated reflective narrative insights of my experiences of change. The methodology was sensitive to context and placed my life and stories at the "forefront of inquiry" (Elbaz-Luwisch, 2012, p. 363). The emergent narrative was independent of "categories of established theory and technique" (Schon, 1983, pp. 68–69). New knowledge was developed and constructed throughout the project's progression. The study did not generate "a set of knowledge claims" (Clandinin & Connelly, 2000, p. 42) that were generalisable (as in empiricist research) to all remote Indigenous secondary mathematics classrooms. Rather, the narrative insights generated offer "a reader a place to imagine their own uses and applications" (Clandinin & Connelly, 2000, p. 42). The study offers readers a place to draw their own implications for effective curriculum and pedagogical practices for Indigenous female middle years (Years 8–9) students, with respect to privileging Indigenous voice and learning measurement within their own personal and social contexts.

1 Narrative: The Experience of Change

The ontological and epistemological grounding of this narrative inquiry was considered along with the four key terms in narrative inquiry: "living", "telling", "retelling", and "reliving" (Clandinin, 2016, p. 34). The narrative inquiry explored the personal, practical, and social elements experienced. The experiences were understood through narrative inquiry commonplaces: temporality, sociality, and place (Clandinin, 2016).

1.1 *Ontological and Epistemological Grounding*

I recorded my experiences of change when contextualising teaching practices to akatyerr harvest through a narrative lens. Through this lens, an experience viewed as a "storied phenomenon" can be (re)storied according to the "social, cultural, familial, linguistic, and institutional narratives" which influence me or can be influenced by me (Clandinin, 2016, p. 33). Narrative inquiry is further understood as a "relational inquiry" (Clandinin, 2016, p. 33). That is, the inquiry is understood in terms of the relational space that exists between the researcher

 | DOI:10.1163/9789004682757_006

and participants in the study. Clandinin and Rosiek (2007) argue that from this narrative view, experience is "more than we can know and represent in a single statement, paragraph or book" (p. 39). Explicitly, they argue "every representation, therefore, no matter how faithful to that which it tries to depict" cannot be used to make generalisable claims since it "involves selective emphasis of our experience" (2007, p. 39). Rather, when experience is considered as that which is "composed and lived over time" (Clandinin, 2016, p. 15), the narrative forms of representation studied are understood "as a series of choices in the whole of an individual or community's live experience" (Clandinin & Rosiek, 2007, p. 40). In the case of this study, I tell of the live practice of decolonising mathematics education in the Utopia Eastern Anmatyerr and Alyawarr community "enacted in storied moments of time and space" (Clandinin & Connelly, 2000, p. 17).

My live experience of teaching and learning in a unique social context – the harvest of wild akatyerr near an outstation within a very remote Northern Territory community – is reported in this book. Here, the term *experience* refers to teaching and learning that took place in both classroom and community contexts. A *reflection in action* approach to this narrative inquiry was appropriate since it generated a story of my experiences as an individual and in relational ways with members of the community (Schon, 1983, pp. 68–69). In particular, I was able to capture my personal and social interactions and explore the meaning of my experiences as told through words, stories, and images in a communicative situation of a "small number of individuals" (Creswell, 2014, p. 508).

1.2 *Four Key Terms*

When viewed through a narrative lens, the narrative phenomenon under investigation can be conveyed as the "living", "telling", "retelling", and "reliving" of experience (Clandinin, 2016, p. 34). These four key terms develop meaning from understanding that the participants and I "live out stories" and "tell stories" of our lived experiences (Clandinin, 2016, p. 34). Further, as I came to live and work alongside Indigenous participants and began to investigate our "lived and told stories", we engaged in "retelling stories" (Clandinin, 2016, p. 34). As we were changed by the lived and told stories of our experience, while retelling stories at the same time, we began to "relive" our stories (Clandinin, 2016, p. 34).

Narrative inquiry can begin with one of two approaches: living stories or telling stories (Clandinin, 2016). Most narrative inquiries begin with the telling (Clandinin, 2016). My study, however, began with our living stories. The participants and I developed meaning from our lived experiences that began while I was living alongside participants (Clandinin, 2016). The narrative inquiry involved grounding pedagogic decision-making in the interests and lived experiences of female First Nations middle years (Years 8–9) students and the aspirations of a First Nations community (Clandinin & Connelly, 2000). The stories were presented through a narrative lens – advocating for "seldom heard" voices in education research (Creswell, 2014, p. 507). I examined my experiences collaborating with First Nations students, assistant teachers, community members, "in relation contextually and temporally" within a remote Australian classroom (Clandinin & Connelly, 2000, p. 189). The purpose of retelling our experiences through this book is to "offer new ways" of designing socially just pedagogy through collaborative decision-making (Clandinin & Connelly, 2000, p. 189).

1.3 *Justifications for Narrative Inquiry: Personal, Practical, and Social*

Narrative inquiry provided a mechanism for Utopia Eastern Anmatyerr and Alyawarr students to have a voice in decisions that impacted their education. The study also supported Utopia Eastern Anmatyerr and Alyawarr assistant teachers to "take on leadership roles" to inform the extent to which their cultural knowledge was embedded in measurement learning (Douglas, 2011, p. 24). For instance, the Indigenous assistant teachers' cultural authority informed the language and akatyerr knowledge concepts shared in yarning circles and later retold in this document. Collaboratively, Indigenous assistant teachers and I shared the decisions regarding the mathematisation of the traditional act, akatyerr wild harvest, within the study design.

The experience of sharing the responsibility for pedagogic decision-making developed mutual respect and trust among the assistant teacher, students, and me (Douglas, 2011). Unlike some forms of qualitative research, which risk "taking without giving back to communities" and which may involve "rushing to quick and possibly ill-conceived design approaches" (Brereton et al., 2014, p. 1183), this study emerged after a prolonged engagement with the community. I lived and worked in community as a teacher for 1 year prior to the conceptualisation of the research. While enrolled as an external Queensland University of Technology (QUT) student part-time, I continued to develop my research focus while engaged in full-time work at the school over the next 3 years. During this time, the Utopia Eastern Anmatyerr and Alyawarr community and I slowly developed meaningful and respectful relationships. The benefits to the Utopia Eastern Anmatyerr and Alyawarr community were developed together over time. The Utopia Eastern Anmatyerr and Alyawarr participants and I "worked through" the pedagogic decision-making in discussion, "waiting for the story to come along" (Christie et al., 2010, p. 72).

The form of narrative inquiry was necessarily ambiguous. A deeper understanding of my experiences emerged during the study as I situated myself in "relational ways" alongside the participants (Clandinin, 2016, p. 34). Narrative insights generated were not able to represent a "timeless authority" (Clandinin & Connelly, 2000, p. 9). Rather, the insights represented the uniqueness of my experiences of the teaching and learning within a unique social context (Clandinin & Connelly, 2000). Insights generated reflected the interaction I had with Utopia Eastern Anmatyerr and Alyawarr participants, the classroom, and the places within community where the study took place. A narrative inquiry framework enabled field texts to be understood in terms of "pieced together patterning", which created "a sense of how things go" (Geertz, 1983, as cited in Clandinin & Connelly, 2000, p. 6) in this unique social context.

1.4 *Narrative Inquiry Commonplaces: Temporality, Sociality, and Place*

Narrative inquiry locates the experience of change in "the context of a specific moment, community and school context, and learning interaction" (Guenther & Osborne, 2020, p. 164). During the study, my role as a participant shifted. At times I was drawn in to "intense involvement", and at other times I stepped back and watched, listened, and recorded observations from a distance (Bateson, 1994, p. 5). Shifting between these roles allowed me to learn while "seeing and telling stories along the way" (Clandinin & Connelly, 2000, p. 9). This book retells my stories of relationships at moments and sites during my teaching in remote Australia. The "ambiguous shifting participant observation role" provided opportunities for me to challenge my assumptions as I met difference (regarding what teaching and learning looks like) "along the way" (Clandinin & Connelly, 2000, p. 9). For instance, my perceptions of what knowledge was included in the curriculum and how it was learned were challenged. I engaged with my bias when critically reflecting on how to effect change in my pedagogic practices. I examined how didactic teaching practices could be transformed to be interactive – within iterative cycles. This document reflects the temporality and situatedness of my role as a participant in the study and as a teacher effecting change in my pedagogy. By grounding pedagogical decision-making in the "aspirations of Indigenous Australians" (Rigney, 1999, p. 114), I resisted the temptation to categorise my understandings of what culturally inclusive measurement curriculum and pedagogic practice (Bateson, 1994, pp. 75–76) look like according to my cultural lens exclusively.

First Nations scholar Martin Nakata (1998) viewed anthropological texts which reported on a historical expedition to the Torres Strait. Nakata's standpoint reveals how, a century earlier, scholars on a colonial settler expedition characterised Torres Strait Islander people as "knowable people" who fit categories which "posit the Islander in a secondary relation to Europeans" (Nakata, 1998, p. 7). I interrogated whether the changes to my pedagogic practice were selected according to a "pre-designated set of categories" derived from a "way of understanding and knowing" which is "discontinuous" with Utopia Eastern Anmatyerr and Alyawarr people's "own order of understanding the world" (Nakata, 1998, p. 7). By referring to the knowledge known to be understood, as depicted in the akatyerr concept map in Figure 12 (Chapter 2), I sought an understanding of the interconnections among akatyerr knowledge interests when and as selected by student participants, their assistant teachers, and community (and/or family) members. Drawing on this knowledge known to be understood was essential to the study. This structured the mathematics explored according to the known relationships of "people", "country", and "creation time" stories (Dobson et al., 2008; Merne Altyerr-ipenhe Reference Group et al., 2011; Turner, 2005).

The decolonisation of mathematics education in this study involved the "recognition of how land and relations to land have always already been differently understood and enacted" (Tuck & Yang, 2012, p. 7). Everyday acts of resurgence (Corntassel, 2012) can reassert Indigenous relationships to land. First Nations within the remote Northern Territory have deep understandings of kinship relations and how they apply to their own observations (ordering and patterning) and understanding of the social and natural world (Lloyd et al., 2016; Verran, 2000, 2008). I reflected on narrative "unities and discontinuities" (Clandinin & Connelly, 2000, p. 17) regarding student and assistant teacher choices in knowledge concepts interrelated to akatyerr to contextualise the measurement learning. Collaborative pedagogical decision-making involved inquiring about the appropriate people to involve in the akatyerr site investigation who were available at the time of the study. The process of investigating akatyerr knowledge concepts to contextualise the measurement learning required respectful and ongoing interactions with participants and the wider Utopia Eastern Anmatyerr and Alyawarr community. The prolonged engagement with community was vital for the study to maintain sensitivity to kinship arrangements with participants and so as not to deny the voices of important members of the community who have a responsibility through kinship to the akatyerr site visited. Privileging Indigenous voice in relation to akatyerr harvest protocols was necessary to collaboratively change the ways of proceeding to include student interest in pedagogic design while maintaining cultural obligations at the akatyerr site visited (see Chapter 4).

2 Understanding the Experiences of Change

Three participant groups were involved in the study: (a) researcher (one female non-Indigenous teacher-researcher – me); (b) adult community members (two female Indigenous assistant teachers and three female and three male Indigenous community members); and (c) students (12 female Indigenous Years 8–9 students). Table 4 provides a rich description of the participants in each group.

Within Table 4 the participants are described according to gender, indigeneity, and role in the community. Additionally, each participant group is described according to temporal, social, cultural, and place considerations. Participant Group 2 included the most experienced assistant teachers at the school, both of whom were also either a parent or grandparent of participating

TABLE 4 Rich description of the participants in each participant group

Participant group	1	2			3
Participants	**Researcher**	**Adult community members**			**Students**
Number, gender, indigeneity and role within the school community	**One female middle years (Years 8–9) non-Indigenous classroom teacher**	**Two female Indigenous assistant teachers**	**Three male Indigenous traditional owners**	**Three female Indigenous traditional owners**	**Twelve female Indigenous middle years students (Years 8–9)**
Temporal (length of service); *social* (learning interactions); *cultural* (kinship); and *place* (remote education setting, school site, akatyerr site).	I had 4 years' experience teaching in a remote education setting and had worked consistently with the assistant teachers and student participant groups for 3 years at the school prior to the akatyerr site visit.	Both had over 10 years' experience teaching at the school and in a remote education setting, and both were related to the akatyerr site through kinship.	All young men were related to the akatyerr site through kinship.	All senior women were related to the akatyerr site through kinship.	All students were available and chose to participate in preliminary yarning circles at the school. Four students who were available and chose to conduct the akatyerr investigation were related to the akatyerr site through kinship.

students. The six adult community members (three female and three male) were related to each other and to the akatyerr site visited through kinship. The 12 female students from a middle years' (Years 8–9) class were recruited because they were willing to participate and were available. The students ranged in age from 12 to 14 years and represented seven family groups from six very remote homeland communities (within a 40-km radius of the school site). Four students, two assistant teachers, three male and two female Indigenous community members, and I were involved in the two recorded yarning circle meetings (2 × 5 minutes) at the akatyerr site.

2.1 *The Recruitment Process*

I chose non-probability sampling with a convenience sampling strategy (Creswell, 2014, p. 144) for participant selection. Non-probability sampling allowed me to select participants in the study who were able to assist with changes to my pedagogic practice because a relational space had developed between the participants and me prior to the study and the participants had an interest in pedagogic design choices that had an impact on the students' learning (Clandinin, 2016, p. 33). A convenience sampling strategy (Creswell, 2014, p. 144) enabled me to seek permission from participants at the research site since I was also employed at the school as a middle years' (Years 8–9) teacher. I approached the assistant teacher participants, who had previously identified their interest in collaborating on changes to my pedagogical practices in measurement to make learning inclusive of Utopia Eastern Anmatyerr and Alyawarr language and culture. Additionally, the research effort ensured there was "no unfair burden" on the student participants, and student wellbeing took precedence over participating in the research (Lassig, 2012, p. 74). Given that Utopia Eastern Anmatyerr and Alyawarr adult participants' voices and images were audiovisually recorded, individual participants or their family members had an opportunity to opt out after they were asked to opt in to the study, if they did not want to continue. In this case, individuals' images were removed from the audiovisual recordings and their anonymity preserved.

2.2 *The Critical Friend and Informed Consent*

The critical friend was selected based on this person being "outside the teaching context" but being able to "provide another perspective to assist in the research effort" (Sagor, 2005, as cited in Foulgar, 2010, p. 148). For the purpose of this study, a linguist was chosen as the critical friend to assist with culturally appropriate ways to interpret field texts within the cultural respect framework (Kendall et al., 2011). Additionally, as the critical friend "did not have a stake in the research problem", he was committed to the research project from a "complementary personal agenda" (Sagar, 2005, as cited in Foulgar, 2010, p. 148).

The critical friend was known to the participants in the study for a prolonged period. He was a trusted and familiar person who assisted with translations at important community meetings involving government officials, at school council meetings, and at local Aboriginal council meetings. He provided a "lens" through which to see the study from another perspective, and in doing so, "elevate" the level and complexity of understanding of the information shared (Costa & Kallick, 1993, p. 50). For instance, the critical friend oversaw the informed-consent process of the study to ensure the informed consent was gained in the participants' first language. He was engaged to assist with explaining the research project and the informed-consent process. The critical friend provided the translations recorded in a short animated video spoken in Eastern Anmatyerr and Alyawarr language. The video, titled *Informed Consent*, elaborated the processes involved in obtaining free and prior informed consent (iTalk Studios, 2018). This video was presented at an informed-consent meeting that enabled informed consent to be obtained in Eastern Anmatyerr and Alyawarr language.

The critical friend and key adult participants ensured the perceptions and critical reflections of the Utopia Eastern Anmatyerr and Alyawarr participants were respectfully and accurately translated and transcribed (Martin, 2008, p. 81). The critical friend responded to this work with "integrity" and as an "advocate" for the success of the study (Costa & Kallick, 1993, p. 50). Further, the critical friend encouraged me to see the dialogue exchanged from "different perspectives" by listening and taking time to understand and clarify the information that was shared in yarning circles (Costa & Kallick, 1993, p. 50).

QUT's processes for obtaining ethical clearance for the study were followed. I obtained prior and informed written consent from all adult participants, parents of students involved in the akatyerr site visit, and student participants. This included explaining the use of all audiovisual material to community members (assistant teachers, traditional owners, and land managers) during the informed-consent meeting, assisted by the interpretation of informed consent in Utopia Eastern Anmatyerr and Alyawarr language in an animation narrated by the critical friend as mentioned above (iTalk Studios, 2018). I explained my position as researcher separate from my role as a teacher. Prior and informed permission was sought so that information obtained by participation in the study could be used to answer the research questions. I approached key members of students' families to respect and show awareness of the importance of the relationships among family members in deciding whether they were interested in their children participating in the study. Additionally, student participants were respected as individuals capable of determining their interest in participating in the research when they were asked for their informed and written consent.

2.3 *Social Context of Teaching and Learning*

The study site was one middle years (Years 8–9) classroom in one high school – Walyarra. As discussed in the previous subsection, the research site is the only high school in this Central Desert region. At the time of this study, approximately 30 students were attending between Year 6 and Year 12 (White & Berta, 2019). The site is an English-only school with an Indigenous Language and Culture program across the school. During 2017–2019, a collaborative teaching partnership developed between Sharmic and me while we delivered the mathematics lessons together. We developed "patterns of interactions" (Scherer & Steinbring, 2006, p. 175) with a cohort of Indigenous female middle years (Years 8–9) students. Established patterns of interactions focused on facilitating students to flexibly move between the "known", or local social and cultural context, to the "unknown", or Western social and cultural context "as represented in the Australian curriculum" (Halsey, 2018, p. 32). For instance, in relation to this study, we facilitated moving between the known when reading the akatyerr-growing landscape as continuous, to the unknown when isolating a discrete area of akatyerr growing within quadrat frames (100 cm × 100 cm). The student cohort was able to experience this pedagogical big idea or notion of turning continuous to discrete through similar approaches to contextualised measurement learning that used the canvases of original artworks to isolate a discrete area (10 mm × 10 mm) (see Chapter 3, Section 1.5).

The established patterns of interactions created opportunities for the same student cohort to "actively produce [their] own knowledge, understanding and explanation" (Scherer & Steinbring, 2006, p. 174) while developing numeracy with measurement attributes such as length and area during their akatyerr investigation. Student participants acted with numeracy as they were able to "visualise, order and describe" the proportion of akatyerr or other vegetation growing within a 100 cm × 100 cm quadrat (ACARA, 2020b). Students applied their proportional reasoning to understand "the relationships of percentages and rate" while solving problems (involving area) in an authentic context (ACARA, 2020a). The student experiences with contextualised measurement learning that occurred prior to the study related the mass of dried akatyerr to price in order to solve problems involving rate such as $/kg. Experiences such as these, which were "composed and lived over time" prior to the akatyerr site investigation, showed students a way to learn measurement through familiar real-life contexts (Clandinin, 2016, p. 15).

The practical justification for "living" my narrative inquiry was "grounded in the need to understand more deeply" my experiences of change when adjusting curriculum and pedagogic practice to be responsive to the students' lived experiences and interests in akatyerr knowledge (Clandinin, 2016, p. 37). I attended more closely to my experience of change so that I "made visible" how teachers might attend differently to decision-making regarding pedagogic design and Australian mathematics curriculum choices to include the lived experiences of students in a remote education setting (Clandinin, 2016, p. 37). I critically reflected on my experiences of change from institutional narrative influences. That is, I engaged in reflecting on the conditions within my role as a teacher at a government school that enabled changes to my existing pedagogic practice (see Table 5).

I used guiding questions to critically reflect on and capture descriptions of adjustments to curriculum and pedagogical practices that progress students' measurement learning. The reflections were audio recorded using iPhone voice memo files, which I sent to my email address and transferred to QUT's research data storage system. These files were then deleted from my iPhone. Reflections captured before and after five yarning circle meetings were guided by the following questions (Mills, 2000, pp. 26–27):

- Who was involved?
- What was it about?
- When did it occur?
- Where did it occur?
- How did it go?

My critical reflections captured "slice of life" (Mills, 2000, p. 74) data at salient points in the narrative inquiry. As I experienced change, when contextualising teaching practices to akatyerr harvest, I saw these experiences through a narrative lens as a "storied phenomenon" (Clandinin, 2016, p. 33). Narrative inquiry, viewed this way, allowed me to examine my experiences from the "social, cultural, familial, linguistic, and institutional narratives" that influence me and can be influenced by me (Clandinin, 2016, p. 33). Table 5 summarises each of these narrative influences on and by me.

The field data were examined through the narrative influences summarised in Table 5. When viewing my field data, I critically reflected on how I designed and lived out my narrative from the personal, practical, and social justifications for an inquiry. For instance, when I critically reflected on my personal justification, I realised it was important to understand how teachers such as myself, with a STEM and equity focus, can experience change when designing socially just pedagogy. For the study, I positioned myself within the school community

TABLE 5 Summary of the influences filtering through my narrative of my lived experience

Influences	Narrative		
	Social	Familiar/linguistic/ cultural	Institutional
Influencing me	I interacted with Indigenous education leaders and scholars such as Chris Sarra, Chris Matthews, Tyson Yunkaporta, and Deborah Cheetham while I worked at an independent secondary girls boarding school for remote and rural Indigenous students. I interacted with many family members from across the Central Desert remote education settings while employed as a classroom teacher with the Northern Territory Department of Education. I sought ongoing guidance from Indigenous academic Chris Matthews for advice on embedding Indigenous perspectives in mathematics education.	Born in Australia within a large Anglo-Celtic family with a low socioeconomic status, I learned to speak English only. I gained a sense of social justice and equity from my family's working-class origins. Through my interest in science, I developed a worldview that aligns with positivist traditions.	I studied a biological science degree specialising in zoology and genetics (conservation of endangered marsupials) and a master's research degree in STEM education and equity at an Australian university. I worked in tertiary and secondary education settings. I joined a not-for-profit organisation, the Aboriginal and Torres Strait Islander Mathematics Alliance (ATSIMA), and presented at two national ATSIMA mathematics education conferences.
Influenced by me	I developed relationships with remote Central Desert communities based on mutual respect and a shared understanding from living and working in a remote location together.	I consider culture as a focal point which I place at the centre of all curriculum I develop. I developed interactive teaching strategies during my experiences teaching students from a wide range of cultural backgrounds as a TESOL teacher (General English and First Certificate of English, Cambridge).	I cultivated a social justice approach within remote Indigenous education. I advocated for student agency and equity to disrupt deficit thinking and associated limiting beliefs in the potential of all Indigenous learners. I gained extensive experience collaborating with community to adapt STEM curriculum and pedagogy in remote education settings.

Note: STEM = science, technology, engineering, mathematics; TESOL = teaching English to speakers of other languages.

as a secondary science and mathematics teacher. This positioning allowed me to use my institutional influences from the STEM education I gained at university to inform some design elements within the narrative inquiry. For instance, landscape and vegetation management teaching techniques, using a transect and quadrat study design, were derived from my familiarity with a positivist tradition. The akatyerr survey design elements, using a randomised trial with a moveable quadrat frame and transect line, permitted students to have agency over their assessments of the akatyerr landscape and vegetation under investigation. That is, students randomly assigned each quadrat point using a random number table; chose the length along the transect line they needed to move to; chose where they needed to place a quadrat frame; and chose the oral, pictorial, and numeric forms of knowledge that would reflect what they observed within the frame.

The relationship of mutual respect between school community members and me created opportunities for exchanging dialogue. These relationships also enabled me to create opportunities for students to record their observations of akatyerr and other vegetation types orally, visually, and numerically. At school community meetings, assistant teachers, parents, and community members had expressed their aspirations for intergenerational knowledge relating to akatyerr to be embedded in school curriculum. I had listened to their aspirations and reflected these in the narrative inquiry design.

The relationships between the assistant teachers, critical friend, and me were crucial to make sense of the contextual aspects of the akatyerr inquiry. For instance, why it was important for students to know certain knowledge concepts in relation to this plant was known from the assistant teachers and community members. The "relational space" among the participants enabled us to develop an understanding of (a) the composition of yarning circle field texts, (b) the critical reflections of my experience of change, and (c) the student artefacts (Clandinin, 2016, p. 34). The assistant teachers, critical friend, and I deepened the understanding of the "composition of the field texts" by interpreting, reading, and member checking the yarning circle field texts together (Clandinin, 2016, p. 34). The multimodal representations of the students' observations provided another opportunity to critically reflect on the telling of the narrative.

When telling the narrative of the akatyerr site visit, for instance, I described how the field data revealed the students' reflexive thinking. I considered the real-life representations of the akatyerr and other vegetation observed by students. The ways the students represented their measurement learning and their use of reflexive thinking were theorised by triangulating my field notes, selected extracts of oral field text, and student artefacts. I theorised how students constructed measurement understanding and knowledge from learning that involved multiple real-life representations. The study permitted students to flexibly move between what they (a) observed within a quadrat frame, (b) captured in a quadrat photograph, (c) drew as a graphical representation, (d) referenced from a guide to estimating percentage cover, (e) symbolised as a number in a table, and (f) exchanged in dialogue. The students' flexibility to move between each form developed their capacity for proportional reasoning.

Chapter 7 presents the theory developed from my study.

References

Australian Curriculum, Assessment and Reporting Authority. (2020a). *Australian curriculum cross-curriculum priorities: Aboriginal and Torres Strait Islander histories and cultures*. Retrieved January 11, 2023, from https://australiancurriculum.edu.au/f-10-curriculum/cross-curriculum-priorities/aboriginal-and-torres-strait-islander-histories-and-cultures/

Australian Curriculum, Assessment and Reporting Authority. (2020b). *Australian curriculum: Learning continuum of numeracy: Using fractions, decimals, percentages, ratios and rates (level 5)*. Retrieved January 11, 2023, from https://bit.ly/3GzjB4i

Bateson, M. C. (1994). *Peripheral visions: Learning along the way*. Harper Collins.

Brereton, M., Roe, P., & Schroeter, R. (2014, April 26–May 1). Beyond ethnography: Engagement and reciprocity as foundations for design research out here. In *CHI'14: Proceedings of the SIGCHI conference on human factors in computing systems, Toronto, ON, Canada* (pp. 1183–1186). https://doi.org/10.1145/2556288.2557374

Christie, M., Guyula, Y., Gotha, K., & Gurruwiwi, D. (2010). The ethics of teaching from country. *Australian Aboriginal Studies*, 2, 69–80. Retrieved January 11, 2023, from https://search.informit.org/doi/abs/10.3316/ielapa.767045875963420

Clandinin, D. J. (2016). *Engaging in narrative inquiry*. Routledge. https://doi.org/10.4324/9781315429618

Clandinin, D. J., & Connelly, F. M. (2000). *Narrative inquiry: Experience and story in qualitative research*. Jossey-Bass.

Clandinin, D. J., & Rosiek, J. (2007). Mapping a landscape of narrative inquiry: Borderland spaces and tensions. In D. J. Clandinin (Ed.), *Handbook of narrative inquiry: Mapping a methodology* (pp. 35–76). Sage Publications.

Corntassel, J. (2012). Re-envisioning resurgence: Indigenous pathways to decolonization and sustainable self-determination. *Decolonization: Indigeneity, Education & Society, 1*(1), 86–101. Retrieved January 11, 2023, from https://jps.library.utoronto.ca/index.php/des/article/view/18627

Costa, A. L., & Kallick, B. (1993). Through the lens of a critical friend. *Educational Leadership, 51*(2), 49–51.

Creswell, J. W. (2014). *Educational research: Planning, conducting, and evaluating quantitative and qualitative research* (5th ed.). Pearson Australia.

Dobson, P. V., Walsh, F., & Sati, W. (2008). *Anpernirrentye: Relationships between bush foods, creation laws, people, country and all things, illustrated by three plants* [DVD]. Desert Knowledge Cooperative Research Centre.

Douglas, J. (2011). *Learning from country: The value of country visits in remote schools for community engagement and natural and cultural resource management* (Desert Knowledge Cooperative Research Centre Report No. 69). Ninti One Limited. Retrieved January 11, 2023, from https://www.nintione.com.au/resource/NintiOneResearchReport_69_Thevalueofcountryvisitsinremoteschools.pdf

Elbaz-Luwisch, F. (2012). Studying teachers' lives and experience: Narrative inquiry into K–12 teaching. In J. D. Clandinin (Ed.), *Handbook of narrative inquiry: Mapping a methodology* (pp. 357–382). SAGE Publications. https://doi.org/10.4135/9781452226552.n14

Foulgar, T. (2010). External conversations: An unexpected discovery about the critical friend in action research inquiries. *Action Research, 8*(2), 132–152. https://doi.org/10.1177/1476750309351354

Guenther, J., & Osborne, S. (2020). Did DI do it? The impact of a program designed to improve literacy for Aboriginal and Torres Strait Islander students in remote schools. *The Australian Journal of Indigenous Education, 49*(2), 163–170. https://doi.org/10.1017/jie.2019.28

Halsey, J. (2018). *Independent review into regional, rural and remote education: Final report.* Commonwealth of Australia. Retrieved January 11, 2023, from https://www.dese.gov.au/quality-schools-package/resources/independent-review-regional-rural-and-remote-education-final-report

iTalk Studios. (2018, November 15). *Informed consent – Alyawarr* [Video]. YouTube. Retrieved December 9, 2022, from https://www.youtube.com/watch?v=30QiWZWCeJc

Kendall, E., Sunderland, N., Barnett, L., Nalder, G., & Matthews, C. (2011). Beyond the rhetoric of participatory research in Indigenous communities: Advances in Australia over the last decade. *Qualitative Health Research, 21*(12), 1719–1728. https://doi.org/10.1177/1049732311418124

Lassig, C. (2012). *Perceiving and pursuing novelty: A grounded theory of adolescent creativity* [PhD thesis, Queensland University of Technology]. QUT ePrints. Retrieved December 9, 2022, from https://eprints.qut.edu.au/50661/

Lloyd, K., Suchet-Pearson, S., Wright, S., Burarrwanga, L., Ganambarr, R., Gunambarr-Stubbs, M., Ganambarr, B., & Maymuru, D. (2016). *Morrku Mangawu—Knowledge on the Land*: Mobilising Yolŋu mathematics from Bawaka, North East Arnhem Land, to reveal the situatedness of all knowledges. *Humanities*, *5*(3), 1–13. https://doi.org/10.3390/h5030061

Martin, K. (2008). *Please knock before you enter: Aboriginal regulation of outsiders and the implications for researchers*. Post Pressed.

Merne Altyerre-ipenhe [Food from the Creation time] Reference Group, Douglas, J., & Walsh, F. (2011). *Aboriginal people, bush foods knowledge and products from central Australia: Ethical guidelines for commercial bush food research, industry and enterprises* (Desert Knowledge Cooperative Research Centre Report No. 71). Ninti One Limited. Retrieved January 11, 2023, from https://www.nintione.com.au/resource/NintiOneResearchReport_71_BushFoodGuidelines.pdf

Mills, G. (2000). *Action research: A guide for the teacher researcher*. Prentice Hall.

Nakata, M. (1998). Anthropological texts and Indigenous standpoints. *Australian Aboriginal Studies*, *2*, 3–12. Retrieved December 9, 2022, from https://search.informit.org/doi/abs/10.3316/aeipt.93483

Rigney, L. I. (1999). Internationalization of an Indigenous anticolonial cultural critique of research methodologies: A guide to Indigenist research methodology and its principles. *Wicazo Sa Review*, *14*(2), 109–121. https://doi.org/10.2307/1409555

Scherer, P., & Steinbring, H. (2006) Noticing children's learning processes – Teachers jointly reflect on their own classroom interaction for improving mathematics. *Journal of Mathematics Teacher Education*, *9*, 157–185. https://doi.org/10.1007/s10857-006-0004-7

Schon, D. A. (1983). *The reflective practitioner: How professionals think in action*. Basic Books.

Tuck, E., & Yang, K. W. (2012). Decolonization is not a metaphor. *Decolonization: Indigeneity, Education & Society*, *1*(1), 1–40. Retrieved December 9, 2022, from https://jps.library.utoronto.ca/index.php/des/article/view/18630

Turner, M. (2005). *Bush foods: Arrernte foods from Central Australia* (2nd ed.). IAD Press.

Verran, H. (2000). Aboriginal Australian mathematics: Disparate mathematics of land ownership. In H. Selin (Ed.), *Mathematics across cultures: The history of non-western mathematics* (pp. 289–311). Springer. https://doi.org/10.1007/978-94-011-4301-1_15

Verran, H. (2008). Mathematics of Yolngu Aboriginal Australians. In H. Selin (Ed.), *Encyclopaedia of the history of science, technology, and medicine in non-Western cultures* (pp. 1432–1437). Springer. https://doi.org/10.1007/978-1-4020-4425-0_8745

White, L., & Berta, L. (2019). *Review of education services in homeland learning centres* [Unpublished draft report]. Northern Territory Government.

CHAPTER 7

Theory of Social Change

Decolonising Mathematics Education

The research was framed as an intervention in my teaching practice. The shifts in my pedagogic practices connected mathematics education to social change. Essentially, the research sought to serve the interests of the wider community to understand how to measure and establish a fairer price for wild akatyerr harvest. What counted as transformation (Smith, 1999) emerged through ongoing consultation with members of community. Students' akatyerr knowledge interests transformed learning to serve the interests of their community. The mathematics education that emerged through the study had an emancipatory dimension. That is, transforming pedagogic practices empowered students to undertake the akatyerr plant survey guided by families and community – through a culturally responsive pedagogy (Morrison et al., 2019). This occurred in yarning circle meetings where "collaborative storying" took place in the form of conversations about students' interests in and lived experiences with akatyerr knowledge concepts (Bishop, 1997). When "the cultural characteristics, experiences, and perspectives" of Utopia Eastern Anmatyerr and Alyawarr students respected "as conduits for teaching them" learning is more effective (Gay, 2002, p. 106). Students' interests and "own experiences" with akatyerr harvest knowledge informed the narrative retold in this book (Bishop & Berryman, 2007, p. 7). The storying gave authority to the voices of the student and adult participants when co-designing the curriculum to be covered during the akatyerr site visit. Community members guided how consideration was given to the students' first language, culture, and interests. The theory presented here describes how my role as a teacher of disciplinary knowledge shifted during the study. The model demonstrates how I acted to interface between the pedagogical adjustments made and the students' expression of their measurement learning.

1 Major Findings

The role of educators in dismantling (or reifying) social inequities has been long known by the giants in critical pedagogy (such as Freire, 1972; hooks,[1] 1994). In Australia, "recognition of Indigenous knowledge systems in our teaching, while challenging inequities in schools" transforms pedagogic practice (Sarra

DOI:10.1163/9789004682757_007

& Ewing, 2021, p. 151). Although not commonly practised, recognising Indigenous knowledge systems as "conduits for teaching" (Gay, 2002, p. 106) allows teachers to re-imagine mathematics learning within and through "cultural knowledges, languages, storytelling, cultural heritage and spiritual connections to land, sea, waterways, and sky" (Sarra & Ewing, 2021, p. 151). I critically examined my role in an Australian public high school mathematics classroom. I have come to know, as a narrative inquirer, how to "create the appropriate conditions for effective learning" (Trueba, 1988, p. 282). That recentring on First Nations language and culture in mathematics learning is "about ways of being in relationship, rather than a focus on subject matter" (Stavrou, 2021, p. 3). As a consequence of the data collection and analysis, I theorised on ways to foreground Indigenous voice to effect changes to measurement curriculum and pedagogic practices in a more inclusive direction. As these changes drew from a community's real-world understanding, the akatyerr investigation facilitated a deeper sharing of knowledge than is possible with a focus on "transmission teaching" of mathematics disciplinary knowledge (Bishop et al., 2009, p. 736). For Utopia Eastern Anmatyerr and Alyawarr people, a familiar social and cultural context (akatyerr harvest) provided opportunities for students to "construct mathematical knowledge" and their "own interpretations of this knowledge, independently" (Scherer & Steinbring, 2006, p. 160). Privileging Indigenous voice (Rigney, 1997, 1999) and decolonising (Chung, 2019, p. 23) student–teacher interaction patterns provided an opportunity for students to engage in a deeper exchange of knowledge and skills relating to sustainable akatyerr harvest. Decolonising actions and interactions "build bridges of meaningfulness between home and school" (Gay, 2000, p. 29). Understanding how the social interaction patterns shifted from didactic to interactive to facilitate two-way dialogue exchanged during the akatyerr quadrat survey is theorised further in the following subsection on the RATIO theory building.

1.1 *RATIO Theory Building*

The RATIO model provides the framework for the theory I have developed here. The theory developed explains how I contextualised measurement learning to the lived realities of students. The RATIO model depicts the following strategies adapted for the akatyerr study: *Reflect*, *Act*, *Trial*, *Interact*, and create *Opportunity*. Each of these strategies builds on the previous. The model includes the trial of the Goompi pedagogical cycle (Matthews, 2006, 2020b), which assisted with the theory building. Figure 40 depicts the framework of the RATIO model used to develop my theory.

The RATIO model scaffolds the retold account of my experiences when contextualising measurement learning. The model encapsulates all the strategies

FIGURE 40 The RATIO pedagogical model phases: Reflect, Act, Trial, Interact, Opportunity

that led to the inclusion of interrelated knowledge concepts of akatyerr of interest to the students and facilitated the new measurement knowledge constructed by the students at the akatyerr site. The model develops a theory to explain how I transformed curriculum and pedagogic practice. The stages – Reflect, Act, Trial, Interact, and create Opportunity – depicted in Figure 40 are elaborated in the following subsections.

1.2 *Reflect*

The RATIO model phase *Reflect* starts with the essential work of all teachers. That is, I needed to reflect on my own worldview and the extent to which my social location as a white, middle class, female teacher influenced my mindset about classroom measurement curriculum and pedagogic practice, my students, and my teaching practices. I critically reflected the measurement disciplinary thinking I was using for area and whether and to what extent the practices I employed to convey this thinking were potentially limiting for my students. For instance, following the recommendations for students

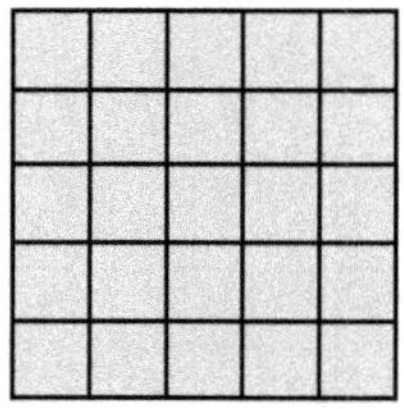

Area = l x w

Area = __________x__________

= ______________square units

FIGURE 41
Example question to demonstrate disciplinary thinking of area of an array (and procedure)
SOURCE: FROM NORTHERN TERRITORY DEPARTMENT OF EDUCATION, 2021, P. 37

in prevocational mathematics (Employment Pathways for Years 7–10) in the Northern Territory, I had asked students to solve the area of an array such as the example in Figure 41 by "emphasis[ing] that it is important students follow all the steps to solve the problem" (NTDOE, 2021, p. 21).

Alternatively, I had asked students to solve the area in a region such as that shown in Figure 7 (Chapter 1). This had provided very limited academic challenge. Students could establish the "area of each shape" by counting the small number of "square units" in green in Figure 7 (NTDOE, 2021, p. 34). I reflected on how examples such as these provided insufficient academic challenge and were not connected meaningfully to the students' lives. Using an array of 5 × 5 cells limited student understanding of area to small whole numbers. I reflected on how students working with a range of numbers (large whole numbers, ratio, percentage, fractions, decimals) were more likely to think flexibly (Siemon et al., 2005). I began to ask myself additional questions, such as: *What harm may I avoid if I adapt my practice?* For instance, I could avoid developing the concept of area in isolation from a big pedagogical idea such as the notion of turning the continuous into the discrete. I considered what I knew of field survey instruments used in ecology, such as transects and quadrats. A quadrat frame designed with a 5 × 5 array could be moved along a transect line so students could experience this big pedagogical idea (isolating a unit of measure) while carrying the quadrat and walking over akatyerr growing (continuous) land.

I considered the students' strengths to guide my reflections. From previous lessons, I knew the student participants could flexibly switch between oral, numeric, and pictorial base representations (Payne & Rathmell, 1977) of fractions, decimals, and percentages. I knew students were able to engage with learning based on country, also from previous lessons. I knew flexibility with a range of numbers was critical to promote students' skill in numeracy. I considered how to connect ideas in measurement, such as area, while promoting fluency so students could flexibly switch between base representations of percentage as a photograph, drawing, symbol, and oral and written numeric form.

I further considered, *What parts of my disciplinary thinking could I adapt?* The use of formulae for constructing area (length × width) could be adapted. For instance, students could solve area by focusing on the patterns of vegetation cover observed within a 5 × 5 quadrat frame. Area, as a percentage of cover, could be solved in a number of ways. For instance, students could focus on the vegetation cover of an individual cell within a quadrat; a number of cells; a row of cells; or a number of cells within each row. They could observe which cells were covered in vegetation and justify the percentage cover over the region of the quadrat they observed in a range of ways. This would allow students to independently recognise and solve area as a direct proportion of the spaces observed with and without vegetation cover. The observations students made could be communicated effectively in a variety of ways (drawings, photographs, oral, symbolic, and numeric forms). Students could engage in constructing knowledge of area while accessing spatial sense, proportional reasoning, and knowledge of place value and number in relation to a real-world problem. That is, students could be prompted to make connections and integrate many mathematical concepts while constructing new measurement knowledge on the presence of akatyerr available for harvest. These reflections on what students knew and their strengths and needs, as well as the boundaries of my disciplinary thinking, were what I perceived as necessary changes to begin to transform my pedagogic practice.

1.3 *Act*

The next phase, *Act*, included the actions and interactions that transformed the teaching patterns from didactic to interactive. This included careful consideration given to which of the students' akatyerr harvest knowledge interests were suitable for the akatyerr investigation. The interaction between Sharmic and me determined the blend of epistemologies that we perceived were suited to the students. Critical to the study was initiating "direct dialogue" (McConaghy, 2000, p. 15) with the assistant teacher in curriculum and pedagogic practice considerations. The change from one-directional learning in the "sender–receiver model" (Ewing, 2011, p. 159) to an interactive model enabled an exchange of ways of knowing, measuring, representing, and retelling interrelated akatyerr harvest concepts within the study. Established social interaction patterns, based on trust and respect, facilitated this collaboration. I regularly checked with Sharmic to decide the "correct rules and ways of proceeding" (Scherer & Steinbring, 2006, p. 174) with knowledge concepts interrelated to akatyerr (creation time, country, and people).

1.4 *Trial*

An important condition for the next phase, *Trial*, was in shifting the dominant focus of communication from the teacher to the learner. Critical also was considering teaching, learning, and measurement knowledge as interdependent. As a result, planned learning occurred with respect to how the teachers, students, and disciplinary knowledge would be expected to interact. For instance, the expectation shifted from the teacher to "know what is correct mathematics" (Scherer & Steinbring, 2006, p. 174) to the students, who were expected to know and be able to represent their own measurement understanding. My role was to provide opportunity for student agency over measurement learning through gradually releasing the communication focus from the didactic to the interactive for this to occur. Figure 42 demonstrates some of the ways the students engaged with the trial of the Goompi pedagogical cycle (Matthews, 2006, 2020b) for the measurement attributes of length and area. The headings in the central panel of Figure 42 – Reality, Abstraction, Mathematics, Reflection – illustrate the core aspects of each phase of the Goompi pedagogical cycle adapted for the trial (Matthews, 2006, 2020b). The outer panels illustrate how each phase of the cycle was adapted for the akatyerr investigation at the site. The image in the top left panel illustrates the reality: the akatyerr site where the yarning circles and quadrat study were conducted. The image in the top right panel illustrates the abstraction phase, where students walked the length along the transect line to place the quadrat frame. The student artefact in the bottom right panel illustrates how the students constructed mathematics knowledge as summarised in the table. The length or distance along the transect was recorded for each quadrat point (in kinaesthetic and numeric forms). The students recorded their spatial and proportional reasoning for the area of the array covered in akatyerr vegetation as a percentage in numeric and symbolised forms (0%). In the bottom left panel, representing the reflection phase, I report on two of the reflexive strategies the students could have used while constructing their learning. The trial planned had embedded opportunities for students to be flexible and generalise about the observations they made. For instance, students could focus on patterns observed in a quadrat frame and could choose to justify their observations in various ways – as photographs, graphs, charts, through two or more oral languages, and as drawings, symbols, numeric, and kinaesthetic representations.

1.5 *Interact*

The *Interact* phase was essential to promote greater student agency. The preceding phases involved ongoing interactions that built iteratively towards greater student agency. For instance, I interacted with my disciplinary thinking in the

Reflect phase. The assistant teacher and I interacted in the Act phase. The assistant teachers, students, and I each interacted with the disciplinary thinking in the Trial phase. Careful considerations of the students, teachers, and ways to interact with disciplinary knowledge provided greater opportunity for the students to interact with disciplinary knowledge on their own terms. The transformation of the didactive to the interactive teaching patterns involved a gradual release of responsibility and authority for interacting with disciplinary thinking from the teachers to the students. The Interact phase led to students autonomously constructing new measurement learning. The interactions between the assistant teacher and me, and between us and the students, combined with our own interactions with the disciplinary thinking, allowed this to occur.

1.6 *Create Opportunities*

Creating meaningful learning *Opportunities* is the last of all the phases described in the RATIO model. Figure 43 depicts an interactive exchange that occurred during a learning opportunity created for the akatyerr investigation. Assistant teacher Sharmic and I both interacted with students while explaining how to construct measurement learning at this akatyerr site. The students, who were sitting on the other side of the quadrat (to the right of this image), recorded their responses at this quadrat point.

The quadrat survey and the yarning circle dialogue captured at the akatyerr site represented the opportunity created for students to "merge" (Lloyd et al., 2016, p. 1) measurement big pedagogical ideas such as "isolating a unit of measure" (YuMi Deadly Centre, 2016) with participants' plant knowledge. Students' knowledge of concepts such as plant names and plant ecology interrelated to akatyerr from lived experiences merged (Lloyd et al., 2016) with scientific inquiry methods (Dobson et al., 2008; Merne Altyerr-ipenhe Reference Group et al, 2011; Turner, 2005). New measurement knowledge was constructed while dialogue engaging each worldview (epistemology) was exchanged.

2 Solutions to Problematic Pedagogic Practices

Problematising my pedagogic practices enabled me to critically reflect on how I had delivered school-based measurement curriculum and what changes I would need to make to transform learning for my students. I examined my role as an educator in a public high school's "cultural meaning-making system" (Gambrell, 2017, p. 2). Pedagogical decision-making that is grounded in a teacher's sense of responsibility can be problematic. The sense of responsibility needs to act as a strong motivator to overcome the burden when engaging in "work that is

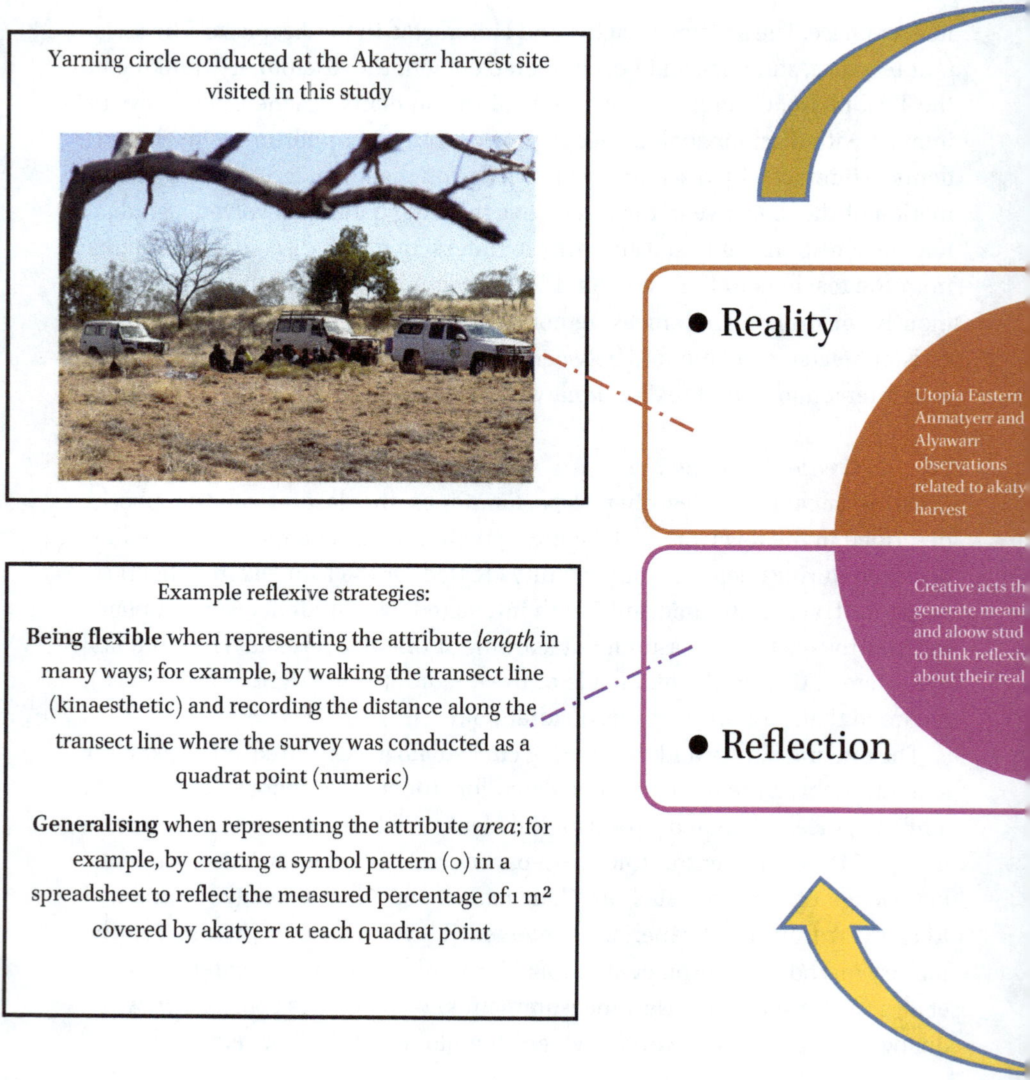

FIGURE 42 Example trial of the Goompi pedagogical cycle – reality, abstraction, mathematics, and reflection (explicitly adapted for the akatyerr quadrat survey; Goompi pedagogical cycle adapted from Matthews, 2006)
PHOTOGRAPHS BY TIM WARE

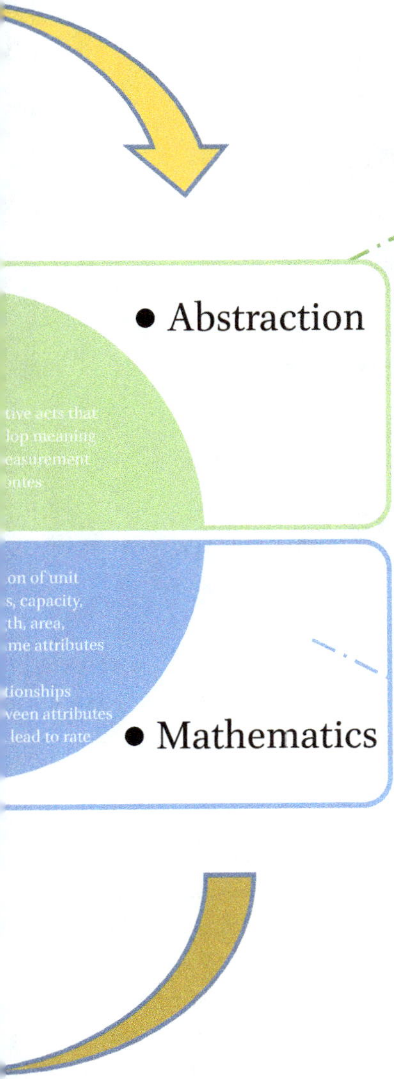

Example creative act performed by students:
Walking the transect line to locate the quadrat point

Example student artefact:

Recording the measurement attributes of *length* (along the transect line to the quadrat point) and *area* (of each quadrat point covered in akatyerr plant as a percentage cover of 1 m^2 quadrat)

Quadrat Study: Akatyerr

Transect line number: 3

Quadrat number	Quadrat point	% Cover
1	37	0
2	13	0
3	50	0
4	27	0
5	45	0
6	9	0

FIGURE 43 Creating opportunities for students to conduct a quadrat survey. Assistant teacher Sharmic is pictured standing, and I am pictured crouching with the quadrat survey instrument held up to demonstrate to students
PHOTOGRAPH BY TIM WARE

not institutionally supported" (Silverman, 2010, p. 10). I investigated how the provision of pedagogic practices grounded in the local context, sustainable akatyerr harvest, enabled student participants to conceptualise and represent measurement ideas in many ways. This involved grounding decisions regarding pedagogical practice in the needs and strengths of my students and their community. Crucially, this involved gradually relinquishing taken-for-granted authority associated with being trained as a teacher (in Western perspectives).

2.1 *Developing an Intercultural Understanding of Ways to Contextualise Measurement Learning*

Yarning circle meetings provided students, assistant teachers, community members, and me with the opportunity for collaborative storying (Bishop, 1997). This gave authority to the voices of the participants contributing to the akatyerr investigation. The use of wild akatyerr harvest presented an opportunity for students to focus on a locally relevant and complex real-world problem, the solutions of which had relevance to the students since they represented the next generation of akatyerr harvesters. The collaborative decision-making supported students to autonomously and creatively construct new measurement learning designed from their interests in an important bush food.

I drew insight from Australian Indigenous mathematician Chris Matthews (2006, 2020b) when considering how I conducted a trial of the Goompi pedagogical cycle in the study. As Matthews (2020a) suggests, a cultural production such as food is underpinned by philosophies, beliefs, traditions, and histories unique to an Australian Indigenous culture. As a non-Indigenous teacher, I cannot know and understand the philosophies underlying the traditional act of harvesting akatyerr. My interpretation of what I saw during an akatyerr harvest learning experience could only be superficial. It was therefore necessary to be guided by the adult participants and seek clarity on the translation of data collected from a critical friend. The critical friend, a linguist known over a prolonged period to the participants, was trusted by participants to assist with translation of the dialogue. A more inclusive curriculum and pedagogic design was created by allowing the voices of community members, knowledgeable in the mathematisation of a traditional act (harvesting akatyerr), to be prioritised. This was especially important when making decisions regarding the conduct of the study while at the akatyerr site.

The pedagogical approach for contextualised learning was also familiar, as it involved learning that was socially constructed at the akatyerr site. Members of the community guided proceedings as they were identified as having cultural responsibility and expert knowledge of that site. International studies in New Zealand serve as a useful parallel to this study. The culturally responsive pedagogical approaches mapped to Māori culture similarly offer alternative pedagogical approaches that challenge deficit discourse of teachers (Bishop et al., 2009). That is, teachers opposed deficit discourse explanations of achievement by committing to and believing that students could attain achievement above previous levels (Bishop et al., 2009). When students know and trust that their teachers believe in their abilities, and show an interest in knowing about their culture, they feel secure. In my study, I established secure attachment in relationships with the students through prolonged engagement as their teacher. Essentially while teaching in ways that respected and cared for the students' wellbeing, I provided the "conditions" where contextualised measurement learning could "most deeply and intimately begin" (hooks, 1994, p. 13).

2.2 *Illustrating Measurement Learning from Students' Interests in Sustainable Wild Akatyerr Harvest*

Altering the pedagogical approach enabled remote female Indigenous students to engage with the Australian Curriculum: Mathematics (Years 8–9), while serving the needs and aspirations of the community for sustaining the local wild akatyerr harvest industry. Shifting pedagogic practices to recentre on this "continents learning system" (rather than a Eurocentric learning system) was decolonising (Rigney, 2017, 45:30). For instance, I recognised "how land and relations to land have always already been differently understood and enacted" (Tuck & Yang, 2012, p. 7). Educators can reconstruct learning systems which respond to the needs of all students – "not limited to one segment of our society" [Christianised, white middle class] (Rigney, 2017, 46:20).

Students learned about their life worlds through two-way dialogue. First, through the local language spoken by elders, then through the mathematical language used to describe and construct their own measurement understanding while solving a real-life problem. Mathematics used in this sense was more about the process of "sharpening our perception" to uncover "patterns that are everywhere around" us (Thurston, 1990, p. 8; see also Matthews, 2020b). The study drew also from students' previous experience on field trips to other akatyerr growing sites. These trips had focused on students' perception of the mass of dried fruit harvested. The study provided opportunities for students to construct new measurement learning in a variety of ways. As established in Chapter 2 (Section 4.3), the ease with which students shift from one representation to another is essential when regrouping and renaming to solve algorithms (Payne & Rathmell, 1977, p. 152). By Year 8, Australian students are expected to flexibly regroup and rename base representations to "solve everyday problems involving rates, ratios and percentages" (ACARA, 2014, p. 17). The focus on vegetation cover reinforced students' earlier learning in the mathematisation of a traditional act to solve an everyday problem. For instance, earlier learning had focused on the price of original locally known artworks solved according to the rate ($\$/cm^2$) and the area of canvas covered in art. Students sharpened their focus on the big pedagogical idea that was common to these contextualised learning experiences. The notion of turning the continuous into the discrete from previous learning experiences permitted students to turn their attention to area of a quadrat frame. Students were able to sharpen their perception when deciding how to visualise, draw, symbolise, and discuss the patterns they observed in the quadrat frame.

3 The Contribution of the Theoretical Frameworks to the Study

The study focused on how the Indigenist principle of privileging Indigenous voice (Rigney, 1997, 1999) supported student agency over their measurement learning. When adult participants engaged in interactive yarning circles, the study was able to "capitalise on every participant's cultural intelligence" and recentre on learning through "the creation of meaning" (Flecha, 2019, p. 662).

3.1 *Adult Participants*

The study foregrounded adult participants' cultural authority to permit students to draw from their own interests and experiences with sustainable akatyerr harvest when constructing new measurement knowledge. The quadrat study facilitated the transfer of cultural knowledge relating to sustainable practices for maintaining wild akatyerr harvest. For instance, during the akatyerr investigation, students were advised by adult participants that they could harvest the yellow ones (ripe fruit) or brown ones (dried fruit) if they were available (harvesting protocols) according to their "own order of understanding the world" (Nakata, 1998, p. 7). The conditions for effective learning were created by recognising the "role of culture" and how it could be used "in the activity settings during the actual learning process" (Trueba, 1988, p. 282).

> At the heart of academic success, and regardless of the child's ethnicity or historical background, an effective learning environment must be constructed in which the child, especially the minority child, is assisted through meaningful and culturally appropriate relationships in the internalisation of the mainstream cultural values embedded in our school system. (Trueba, 1988, p. 282)

Meaningful and culturally appropriate relationships occurred with and through the communicative situation and interactions, primarily with assistant teachers.

3.2 *Assistant Teachers*

Intuitively, adult akatyerr harvesters know and understand the capacity of an akatyerr site for yield. Like an expert mathematician, assistant teacher and wild akatyerr harvester Sharmic translated one mental mode (the area of a quadrat) to another mental mode (the volume of a bucket) "almost unconsciously" (Thurston, 1994, p. 6). Sharmic held the logical structures for measuring area and volume in her head, while she made a statement of her proportional reasoning. She predicted that, at the akatyerr site, students may harvest "one bucket or two bucket or even three" of dried akatyerr for each square metre. Her statement reflected her personal thinking, which, like expert mathematicians, would be difficult to convey to an audience who were "not familiar with this thinking" (Thurston, 1994, p. 14). Sharmic's cultural authority, combined with her experience and understanding of school-based mathematics, made a significant contribution to the students' learning. The trial of a Goompi pedagogical cycle (Matthews, 2006, 2020b) relied on Sharmic's expert knowledge. Students could critically reflect on their observations during the akatyerr investigation, while guided in discussions by the observations known to experienced akatyerr harvesters. Sharmic's significant contributions supported students to access new ways of constructing measurement learning at the akatyerr site.

3.3 *Students*

The study supported the whole learner by listening to what the students were interested in with regard to wild akatyerr harvest knowledge concepts and measurement concepts known by the community. The study allowed the students to represent their understanding of length and area in multiple ways during the quadrat and yarning circle study. The learner was free to "express their authentic selves without fear of being negatively judged" (Giorcelli, 2020, slide 23) using their home and school-based linguistic and cultural knowledge. Students were provided opportunities to communicate the mathematical connections they made at the akatyerr site using a range of skills: visual, kinaesthetic, and spatial senses, as well as language, symbols, images, graphs, and photographs. US mathematician William Thurston suggested that any mathematical concept is best represented as an "integrated combination of several different representations" (as cited in Davis, 2020, p. 1). Illustrating the measurement attributes of area and length, using multiple ways, was more likely to allow students to engage their "full intellect" (Thurston, 1990, p. 9). The mental image of the plant within the quadrat was held in their mind while drawing the visual representation of the vegetation cover. Students related to what they were learning by holding "logical structures" in their head while engaging with situations that were "real to them" (Thurston, 1990, p. 9).

3.4 *Bush Food Industry Experts*

At the akatyerr site, as Rayleen Brown has cited for other areas, there was a need for sustainable land management practices since there is no ranger program protecting it. As Rayleen buys akatyerr from the women, she has been listening to their concerns and has advocated for ranger programs to support the harvesters' needs. This study permitted the young female middle years students, who are the next generation of harvesters, to record knowledge concepts interrelated to akatyerr as presented by the men and women responsible for this site.

3.5 *Experts Grounding Pedagogical Decisions in Social Justice*

Adapting the Goompi model (Matthews, 2006, 2020b) to begin with the students' real-life experiences and interests in akatyerr, and privileging and integrating Indigenous voices in the curriculum, created the opportunity for akatyerr knowledge concepts to be investigated. Moreover, assistant teachers such as Louisa provided opportunities for students to access "traditional culture and acts" in relation to "plant knowledge" in a way such that this right was protected (Janke & Cubillo, 2019, p. 1). Explicitly, the students (young women) could harvest akatyerr, a "right" they have to "traditional culture and acts" (Janke & Cubillo, 2019, p. 1), while also constructing middle years measurement knowledge relating to their observations of akatyerr at the site.

3.6 *The Critical Friend*

The important work of this study could not have been achieved without the non-partisan interpreter or critical friend. The critical friend acted as a cultural conduit, moving through liminal spaces (passages) between the culture and language understandings of the Utopia Eastern Anmatyerr and Alyawarr participants and the culture and language understandings embedded within a remote education setting (Hogue, 2019). As I am a non-Indigenous outsider, it was of critical importance that I embedded the work of the cultural navigator to effectively represent the voice of the participants in the study. The critical friend represents one of the key contributions this study makes to the scholarship of teaching and learning in both-ways education (Ober, 2009; Ober & Bat, 2007).

4 The Contribution of the Thematic Analysis

The students asked the adult participants about knowledge concepts related to sustainably harvesting akatyerr, an important plant food for Utopia Eastern Anmatyerr and Alyawarr people. Each learner's identity, as a female middle years and Indigenous mathematics student, was framed during the study from a strengths discourse. Knowing about the harvesting protocols, rainfall, plant names, animals, and sustainable management of wild akatyerr were identified as important to the student participants. Learning, designed for and from each learner's strengths in akatyerr knowledge concepts such as these, enabled students to take greater agency from their identified interests in plant identification, for instance, while engaging their diverse oral linguistic and spatial skills.

The careful consideration given to the right people, place, and time when visiting the akatyerr site ensured "traditional acts" were protected (Janke & Cubillo, 2019, p.1), and the research participants did not accidentally act the wrong way. Kinship considerations supported the right for students to participate in traditional acts such as harvesting akatyerr while on site. Rayleen Brown, a prominent Ngangiwumirr/Arrernte bush food industry spokeswoman, foregrounded the need to respond to the different responsibilities men and woman have to akatyerr growing land according to kinship, and the need for Indigenous land management approaches to support sustainable harvest of bush foods, in an interview cited in Finnane (2018):

> In the areas where the bush foods are being harvested, they don't have ranger programs, so that's a difficulty in itself. The ladies want all their young men to do that, it is men's work to do the burning. (para. 34)

The young men participating in the study shared with the students their knowledge of where and how the site would be burned. Similarly, the women harvesters shared their knowledge with the students about sustainable ways to manage the akatyerr harvest site.

5 Limitations of the Study

The akatyerr investigation culminated in one half-day visit to an akatyerr harvest site. The investigation involved a class comprised of 12 female students in Years 8 and 9. All spoke two or more Indigenous languages and English as an additional language. A gender perspective was not considered in the study. However, gender in STEM education is acknowledged as a significant barrier to equitable access. Industry experts from the Australian Academy of Science (2019) recognise the need to create relevance to female students in education to address inequity that relates specifically to gender in STEM careers. The "gender gap in STEM fields is wide" because female students have been more significantly impacted by "limiting teaching practices" than their male counterparts (Carrillo, 2019, p. 2). Although gender inequity is in urgent need of redress, my study revealed issues of power most urgently in relation to racially based forms of oppression, specifically in Australian Indigenous education within remote Northern Territory regions (Fogarty et al., 2015).

The study focused on improving access to curriculum and pedagogic practice through reflective practice challenging unconscious bias that may inadvertently oppress Indigenous students. Acting with cultural integrity by privileging the voices of the student and adult participants promoted a more equitable and integrated experience of STEM education than prior to the study. Moreover, it provided me with an opportunity to oppose inauthentic pedagogy and develop my racial consciousness. That is, I learned a "deeper awareness and understanding of race" (AITSL, 2020, p. 19). Through experiences co-creating authentic lessons with assistant teachers, I developed an appreciation for how to promote student agency and equitable access for Indigenous female students to the Australian Curriculum: Mathematics (Years 8–9).

The study was limited to the measurement attributes of area and length. The students were unable to make further and continued connections with other measurement attributes (mass, capacity, and volume). This was because at the time of the akatyerr site visit (June 13, 2019), the akatyerr plants present had not produced fruit. Further opportunity for two-way dialogue on rainfall observations could have led to the akatyerr growing landscape being contextualised to volume. For instance, altering the length of a transect overlaying a harvest site could have created opportunities for students to investigate the relationship between measures of length, area, and volume with respect to 1 mm of rainfall on the 1 m^2 quadrat site. Additionally, further opportunities could have facilitated direct dialogue on the student representations of 1 m^2 area (photograph image) to elicit measures of volume of rain falling on the akatyerr harvest site (see Chapter 8, Figure 46). Also, measurements of the rate of yield harvested (g/m^2) and price ($\$/m^2$) according to the length of the land (50 m transect) could have been calculated. These calculations in relation to 1 m^2 could have been applied to a larger akatyerr growing area to estimate the economic value of the chosen harvest site. Also, a measure of the area parallel to the site could have been calculated using parallel transect lines defined on an aerial map (see Chapter 8, Figures 47 and 48).

6 Recommendations for Mathematics Education

The overarching recommendation for mathematics education is to frame the learner within a strengths discourse and positive perception of potentiality (Fogarty et al., 2015). The quality of mathematics pedagogy in this study improved by disrupting my perception of students' potentiality and grounding my pedagogic decision-making in the aspirations of the students, their families, and community members. A student cohort can deepen their mathematics content knowledge and develop capabilities from a real-life and meaningful context when teachers tune in to the needs and strengths of the students.

I recommend that mathematics educators consult directly with local Indigenous cultural advisors. The contextualised measurement curriculum in this study was broadly developed in consultation with Central Desert harvesters, traders, primary industry partners, anthropologists, land council officers, local Indigenous board members, Indigenous assistant teachers, and non-Indigenous teachers. The study focused on the sustainable wild harvest of akatyerr since learning about the economic development of this bush food supports students to develop "entrepreneurial skills" (Education Council, 2019, p. 8). "Entrepreneurial education requires schools to reach out into their communities and beyond to explore and engage with 'real world' possibilities" (Halsey, 2018, p. 20).

Additionally, I recommend community participation to develop contextualised curriculum that is relevant to the needs and aspirations of the learner and the wider community. There is increased need for "direct participation" of remote "educators, students, parents/families, employers" during the development of Australian Curriculum and assessments (Halsey, 2018, p. 33). This is to ensure the Australian Curriculum and pedagogical practices are relevant for very remote students and their communities. Similarly in Aotearoa (New Zealand), supporting equitable learning outcomes involves "ncourage[ing] collaborations across diverse perspectives to support authentic learning partnerships with students, parents, whānau, iwi [tribes], and communities through power sharing" (Curriculum, Progress, and Achievement Ministerial Advisory Group, 2019, p. 60).

Finally, I recommend privileging the voice of participants through the act of asking whether participants wish their identity (language and culture) to be recognised in Indigenous research. Researchers should ask themselves, *How much anonymisation should occur?* In order to fulfil traditional institutional research imperatives for ethical research to require anonymity of participants – must we erase Indigenous identity? Should we remove the right to share the identity of Indigenous participants if we seek to privilege Indigenous voice and identity?

7 Directions for Further Research

This study assisted the local bush food industry through supporting intergenerational transfer of akatyerr harvesting knowledge and skills (CLC, 2017, 2018). These skills help prepare students to actively participate in an "increasingly innovation-based economy" (Ewing, 2011, p. 75). Further research could collaboratively develop "entrepreneurial skills" (Education Council, 2019, p. 8). This will involve shifting teaching from a didactic to a student-centred approach. Additionally, developing entrepreneurial skills will require "a set of big ideas and questions" that permit mathematics and other disciplinary learning to be "the basis of inquiry" in which unique solutions are sought rather than knowledge being seen as "the basis of learning" (Halsey, 2018, p. 68).

Further research should focus on facilitating students to develop mathematical problem-solving skills that serve the needs and aspirations of their local communities. The students who participated in the akatyerr survey now have experience in recording and monitoring plants and stories at this important akatyerr site. Indigenous ecological knowledge (IEK) is knowledge of sustainable land management practices that Indigenous Australians have acquired "over thousands of years" and "passed through generations" (Janke, 2021, p. 236).

Despite the site being identified for its significant IEK value by Utopia Eastern Anmatyerr and Alyawarr people, students have little or no future opportunities in local ranger employment to build on this knowledge. At the time of writing this document, the Australian Government had committed to providing money to support vocational training and skills in conservation and land management for this community. In the future, the contextualised measurement learning developed in this project could be applied to the emerging ranger programs for this and other akatyerr-growing desert communities. Chapter 8 elaborates how the site selected for monitoring akatyerr can recentre on the aspirations of the next generation and the emerging ranger programs, starting with culturally responsive pedagogy for Year 8 mathematics curriculum and moving towards vocational education and training (VET) qualifications that can be applied in other ecosystems (marine, for example).

Note

1 By not capitalising the first letters of her name, belle hooks challenges grammar prescriptivism. As a construct, language can facilitate taken-for-granted practices of racism and gender bias.

References

Australian Academy of Science. (2019). *Women in STEM decadal plan.* Retrieved January 11, 2023, from https://www.science.org.au/womeninSTEMplan

Australian Curriculum, Assessment and Reporting Authority. (2014). *Australian curriculum: Work sample portfolio – Mathematics year 8: Above satisfactory.* Retrieved January 11, 2023, from https://docs.acara.edu.au/curriculum/worksamples/Year_8_Mathematics_Portfolio_Above.pdf

Australian Institute of Teaching and School Leadership. (2020). *Indigenous cultural competency in the Australian teaching workforce: Discussion paper.* Retrieved January 11, 2023, from https://www.aitsl.edu.au/docs/default-source/comms/cultural-competency/aitsl_indigenous-cultural-competency_discussion-paper_2020.pdf

Bishop, R. (1997). Interviewing as collaborative storying. *Education Research and Perspectives, 24*(1), 28–47. Retrieved January 11, 2023, from https://search.informit.org/doi/10.3316/aeipt.79388

Bishop, R., & Berryman, M. (2007). *Culture speaks: Cultural relationships and classroom learning.* Huia Publishers.

Bishop, R., Berryman, M., Cavanagh, T., & Teddy, L. (2009). Te Kotahitanga: Addressing educational disparities facing Māori students in New Zealand. *Teaching and Teacher Education, 25*(5), 734–742. https://doi.org/10.1016/j.tate.2009.01.009

Carrillo, J. (2019). *Mindset, math and gender: How impacting female students' mindset towards math impacts their success in the classroom and beyond* [Doctoral dissertation, California State University, Fullerton]. ProQuest Dissertations Publishing. Retrieved January 11, 2023, from https://www.proquest.com/openview/0851aa7678f8d47d227f7081d7c34bed/1?pq-origsite=gscholar&cbl=2026366&diss=y

Central Land Council. (2017). *Central Land Council annual report 2016–17*. Retrieved December 9, 2022, from https://www.clc.org.au/wp-content/uploads/2021/03/CLC-Annual-Report-2016-2017.pdf

Central Land Council. (2018, May). Growing the growers – when bush harvesters meet markets. *Land Rights News Central Australia, 8*(1), 21. Retrieved December 9, 2022, from https://www.clc.org.au/land-rights-news/

Chung, S. (2019). The courage to be altered: Indigenist decolonization for teachers. *New Directions for Teaching and Learning, 2019*(157), 13–25. https://doi.org/10.1002/tl.20327

Curriculum, Progress, and Achievement Ministerial Advisory Group. (2019). *Strengthening curriculum, progress, and achievement in a system that learns. E whakakaha ana te marautanga, te koke, me te ekenga taumata te rangapū e ako ana*. Retrieved January 7, 2023, from https://conversation-space.s3-ap-southeast-2.amazonaws.com/ELS+0324+CPA+Final+MAG+report_06+includes+Ed+Strategy+vision.pdf

Davis, D. (Ed.). (2020). *Illustrating mathematics*. American Mathematical Society. Retrieved December 9, 2022, from https://dianadavis.github.io/im/book.pdf

Dobson, P. V., Walsh, F., & Sati, W. (2008). *Anpernirrentye: Relationships between bush foods, creation laws, people, country and all things, illustrated by three plants* [DVD]. Desert Knowledge Cooperative Research Centre.

Education Council. (2019). *Alice Springs (Mparntwe) education declaration*. Retrieved January 11, 2023, from https://nla.gov.au/nla.obj-2391375250/view

Ewing, B. (2011). Direct instruction in mathematics: Issues for schools with high Indigenous enrolments: A literature review. *Australian Journal of Teacher Education, 36*(5), 65–92. https://doi.org/10.14221/ajte.2011v36n5.5

Finnane, K. (2018, March 2). Bush foods: How can wild harvesters get a piece of the pie? *Alice Springs News, 25*(2). Retrieved January 11, 2023, from https://alicespringsnews.com.au/2018/03/02/bush-foods-how-can-wild-harvesters-get-a-piece-of-the-pie/

Flecha, R. (2019). Exploring the impact of interactive groups: Dialogic interactions involving families and community members. In N. Mercer, R. Wegerif, & L. Major (Eds.), *The Routledge international handbook of research on dialogic education*. Routledge. https://doi.org/10.4324/9780429441677

Fogarty, W., Lovell, M., & Dodson, M. (2015). A view beyond review: Challenging assumptions in Indigenous education development. *UNESCO Observatory Multi-Disciplinary Journal in the Arts, 4*(2), 1–23. Retrieved October 14, 2022, from https://hdl.handle.net/1885/13420

Freire, P. (1972). *Pedagogy of the oppressed.* Penguin.

Gambrell, J. A. (2017). Is culturally responsive pedagogy enough? Toward culturally "real"-evant curriculum. A response to "Democratic Foundations for Spiritually Responsive Pedagogy". *Democracy & Education, 25*(1), Article 12. Retrieved December 14, 2022, from https://democracyeducationjournal.org/home/vol25/iss1/12

Gay, G. (2000). *Culturally responsive teaching: Theory, research, and practice.* Teachers College Press/Columbia University.

Gay, G. (2002). Preparing for culturally responsive teaching. *Journal of Teacher Education, 53*(2), 106–116. https://doi.org/10.1177/0022487102053002003

Giorcelli, L. (2020, November 23). *Keynote presentation: Day 2 – Dr Loretta Giorcelli* [Video]. Northern Territory Department of Education Leaders' summit 2020: Beyond the horizon. Retrieved January 11, 2023, from https://www.youtube.com/watch?v=uEHMNXbOHZ0

Halsey, J. (2018). *Independent review into regional, rural and remote education: Final report.* Commonwealth of Australia. Retrieved January 11, 2023, from https://www.dese.gov.au/quality-schools-package/resources/independent-review-regional-rural-and-remote-education-final-report

Hogue, M. (2019). Two-eyed seeing for both ways knowing: Using digital technology to create liminal spaces for Indigenous STEM success. *Discussions on university science teaching: Proceedings of the Western Conference on Science Education, 2*(1). Retrieved January 11, 2023, from https://ojs.lib.uwo.ca/index.php/wcsedust/article/view/8110

hooks, b. (1994). *Teaching to transgress: Education as the practice of freedom.* Routledge.

Janke, T. (2021). *True tracks: Respecting Indigenous knowledge and culture.* NewSouth Publishing.

Janke, T., & Cubillo, C. (2019). *Towards an ICIP (Indigenous cultural and intellectual property) strategy for your work at Walyarra School* [pseudonym]. Terri Janke and Company.

Lloyd, K., Suchet-Pearson, S., Wright, S., Burarrwanga, L., Ganambarr, R., Gunambarr-Stubbs, M., Ganambarr, B., & Maymuru, D. (2016). *Morrku Mangawu—Knowledge on the Land*: Mobilising Yolŋu mathematics from Bawaka, North East Arnhem Land, to reveal the situatedness of all knowledges. *Humanities, 5*(3), 1–13. https://doi.org/10.3390/h5030061

Matthews, C. (2006). *Care for Country initiative* [Unpublished manuscript]. Griffith University.

Matthews, C. (Guest Host). (2020a). Making a connection to culture (No. 38) [Podcast episode]. In *The random sample*. ARC Centre of Excellence for Mathematical and Statistical Frontiers (ACEMS). Retrieved December 15, 2020, from https://acems.org.au/podcast/episode-38-chris-matthews

Matthews, C. (2020b). *Teaching mathematics from a cultural perspective*. State of Victoria (Department of Education and Training). Retrieved December 19, 2022, from https://www.education.vic.gov.au/school/teachers/teachingresources/discipline/maths/Pages/research_culturalperspective.aspx

McConaghy, C. (2000). *Rethinking Indigenous education: Culturalism, colonialism, and the politics of knowing*. Post Pressed.

Merne Altyerre-ipenhe [Food from the Creation time] Reference Group, Douglas, J., & Walsh, F. (2011). *Aboriginal people, bush foods knowledge and products from central Australia: Ethical guidelines for commercial bush food research, industry and enterprises* (Desert Knowledge Cooperative Research Centre Report No. 71). Ninti One Limited. Retrieved January 11, 2023, from https://www.nintione.com.au/resource/NintiOneResearchReport_71_BushFoodGuidelines.pdf

Morrison, A., Rigney, L., Hattam, R., Diplock, A. (2019). *Toward an Australian culturally responsive pedagogy: A narrative review of the literature*. University of South Australia. Retrieved December 9, 2022, from https://apo.org.au/node/262951

Nakata, M. (1998). Anthropological texts and Indigenous standpoints. *Australian Aboriginal Studies*, *2*, 3–12. Retrieved December 9, 2022, from https://search.informit.org/doi/abs/10.3316/aeipt.93483

Northern Territory Department of Education. (2021). *Indigenous education strategy employment pathways: Communication and problem solving: Numeracy – Area (Version 2.2)*. Northern Territory Government.

Ober, R. (2009). Both-ways: Learning from yesterday, celebrating today, strengthening tomorrow. *Australian Journal of Indigenous Education*, *38*(Supplement), 34–39. https://doi.org/10.1375/S1326011100000806

Ober, R., & Bat, M. (2007). Paper 1: Both-ways: The philosophy. *Ngoonjook: A Journal of Australian Indigenous Issues*, *31*, 64–86. Retrieved January 11, 2023, from https://eprints.batchelor.edu.au/id/eprint/99/1/Ngoonjook31_Ober&Bat.pdf

Payne, J., & Rathmell, E. (1977). Numeration. In J. Payne (Ed.), *Mathematics learning in early childhood: 1977 yearbook of the National Council of Teachers of Mathematics*. NCTM.

Rigney, L. I. (1997). *Racism and physical education: A critical Indigenist analysis of the Senate Standing Committee's report on physical and sport education* [Unpublished master's thesis]. University of South Australia.

Rigney, L. I. (1999). Internationalization of an Indigenous anticolonial cultural critique of research methodologies: A guide to Indigenist research methodology and its principles. *Wicazo Sa Review*, *14*(2), 109–121. https://doi.org/10.2307/1409555

Rigney, L. I. (2017, September 29). Decolonising Pacific schools: Toward an Australian culturally-responsive pedagogy? [Video]. YouTube. Retrieved December 8, 2022, from https://www.youtube.com/watch?v=bpqcU27kkWc

Sarra, G., & Ewing, B. (2021). Culturally responsive pedagogies and perspectives in mathematics. In M. Shay & R. Oliver (Eds.), *Indigenous education in Australia: Learning and teaching for deadly futures* (pp. 148–161). Routledge. https://doi.org/10.4324/9780429263453

Scherer, P., & Steinbring, H. (2006) Noticing children's learning processes – Teachers jointly reflect on their own classroom interaction for improving mathematics. *Journal of Mathematics Teacher Education, 9*, 157–185. https://doi.org/10.1007/s10857-006-0004-7

Siemon, D., Breed, M., & Virgona, J. (2005). From additive to multiplicative thinking: The big challenge of the middle years. In J. Mousley, L Bragg, & C. Campbell (Eds.), *Mathematics – celebrating achievement: Proceedings of the 42nd conference of the Mathematical Association of Victoria*. MAV.

Silverman, S. K. (2010). What is diversity? An inquiry into preservice teacher beliefs. *American Educational Research Journal, 47*(2), 292–329. https://doi.org/10.3102/0002831210365096

Smith, L. T. (1999). *Decolonizing methodologies: Research and Indigenous peoples*. Zed Books.

Stavrou, S. G. (2021). A narrative inquiry into indigenizing school mathematics through miyō-pimōhtēwin and kamskénow. *Frontiers in Education, 6*. https://doi.org/10.3389/feduc.2021.679087

Thurston, W. (1990). Mathematical education. *Notices of the American Mathematical Society, 37*(7), 844–850. Retrieved December 9, 2022, from https://arxiv.org/pdf/math/0503081v1.pdf

Thurston, W. (1994). On proof and progress in mathematics. *Bulletin of the American Mathematical Society, 30*(2), 161–177. https://doi.org/10.1090/S0273-0979-1994-00502-6

Trueba, H. T. (1988). Culturally based explanations of minority students' academic achievement. *Anthropology & Education Quarterly, 19*(3), 270–287. https://doi.org/10.1525/aeq.1988.19.3.05x1565e

Tuck, E., & Yang, K. W. (2012). Decolonization is not a metaphor. *Decolonization: Indigeneity, Education & Society, 1*(1), 1–40. Retrieved December 9, 2022, from https://jps.library.utoronto.ca/index.php/des/article/view/18630

Turner, M. (2005). *Bush foods: Arrernte foods from Central Australia* (2nd ed.). IAD Press.

YuMi Deadly Centre. (2016). *YuMi Deadly Maths: Measurement – Prep to Year 9*. Queensland University of Technology. Retrieved December 9, 2022, from https://research.qut.edu.au/ydc/wp-content/uploads/sites/181/2019/12/YDM-Book6-Measurement-2016.pdf

CHAPTER 8

Future Generations

The young Aboriginal people involved in this study are from a generation that has "far fewer opportunities" to learn the many and complex knowledge concepts interconnected to significant bush food species like akatyerr (Walsh et al., 2013, p. 14). Country, People, and Creation Time knowledge concepts represented in Figure 12 (Chapter 2) "held by senior Aboriginal people who grew up reliant on bushfoods and other resources" were made visible to the students participating in the study using the conceptual map (Walsh et al., 2013, p. 14). The akatyerr plant survey contributed new knowledge about how students could construct new measurement learning through a "connection to place, story, and identity" (First Nations Education Steering Committee & First Nations Schools Association, 2020, p. 22).

1 Anti-Colonial Educational Perspectives for Transformative Change

The collective needs, aspirations, and perceptions of First Nations peoples in Canada and the USA rarely inform policy measures designed to address "socioeconomic gaps with the mainstream" (Cornell & Jorgensen, 2019, p. 287). Similarly, in Australia, instead of focusing on "individualized socioeconomic metrics" informed by "close-the-gap thinking", the challenge is to address the "assumption that the mainstream norm should be the policy goal" (Cornell & Jorgensen, 2019, p. 287). For instance, considering the question, "What are the collective needs of Central Australian Desert First Nations?" then working with these nations to address these needs. Unlike First Nations in Canada, Australian anthropologists involved in "land claims and native title proceedings" are focused on "Aboriginal people's testimony about their connections to land through sacred sites, spiritual associations, and social relationships" rather than Aboriginal uses of resources (Walsh et al., 2013, p. 11). How do we recentre on the collective needs of First Nations to protect plant resources in Australia?

In Amazonian Ecuador, "young Waorani adults" are engaged in "technical training and employment" in the use of "environmental mapping and monitoring" to document their extensive knowledge of Amazonian forest plant resources (High, 2020, p. 308). The maps have aided decisions to prevent extractive economies (oil companies) – since they can demonstrate the vast Waorani knowledge of plants and reliance on their uses. For instance,

 | DOI:10.1163/9789004682757_008

technical field training allowed Waorani to map GPS data points during walks through significant cultural trails. GPS data points were recorded into "Mapeo ..., exported from Mapeo to geojson, and then uploaded into Mapbox" (Digital Democracy, n.d., paras. 6–7). In this way, "Waorani history, cultural and ethnobotanical information" could be overlaid in satellite imagery which captures changes in "elevation" or "forest cover", for example (Digital Democracy, n.d., para. 6). However, during native title proceedings in Australia, "Aboriginal uses of resource species have largely been ignored" (Walsh et al., 2013, p. 11). In this study, one significant plant species in one Central Desert community was foregrounded. The adults' dialogue collected advised the protocols to facilitate deep understanding of akatyerr cultural knowledge shared. The interconnections – for instance, between creation stories of akatyerr and people – teach important moral codes (Walsh et al., 2013, p. 11). For example, senior Warlpiri knowledge holders responsible for invoking moral codes embedded in dreaming stories have related landscape features or "granitic boulders" to ancestral bird species, namely, the emu and bustard. The two ancestral characters or two bird species responsible created "balls of dried" akatyerr to teach about "greed and deceit" when fighting over this significant resource (Walsh et al., 2013, p. 10). Deep knowledge embedded in creation stories is also shared across neighbouring communities. For instance, the map, titled "Bandaiyan", meaning a "body into which the law of the *Wunnan* sharing system is woven" (Morris, 2007, p. 13), was created by senior lawman David Mowaljarlai of the Ngarinyin People of north-west Australia's Kimberley region. Senior lawman Mowaljarlai produced "a 'template' for 'sharing' the management of Bandaiyan – Corpus Australis" (Morris, 2007, p. 108). The template, drawn with the assistance of a biographer (Sutton, 1998, p. 415) documents "dreaming songlines" (Walsh et al., 2013, p. 8). The map depicts the "vast areas" where "ceremonies and knowledge exchange" maintain interconnections across Australia (Walsh et al., 2013, p. 8). Senior lawman Mowaljarlai shares,

> I want to show you something. I want to show you how all Aboriginal people in Australia are connected in the Wunnan system. The squares are the areas where the communities are represented, and their symbols and the languages of the different tribes in this country from long-long time ago. The lines are the way the history stories travelled along these trade routes. They are all interconnected. It's the pattern of the Sharing system. (Mowaljarlai & Malnic, 1993, p. 190, as cited in Morris, 2007, p. 96)

Similarly, neighbouring language groups connected to Alyawarr and Eastern Anmatyerr, such as Warlpiri or Kaytetye (see Figure 44), maintain connections

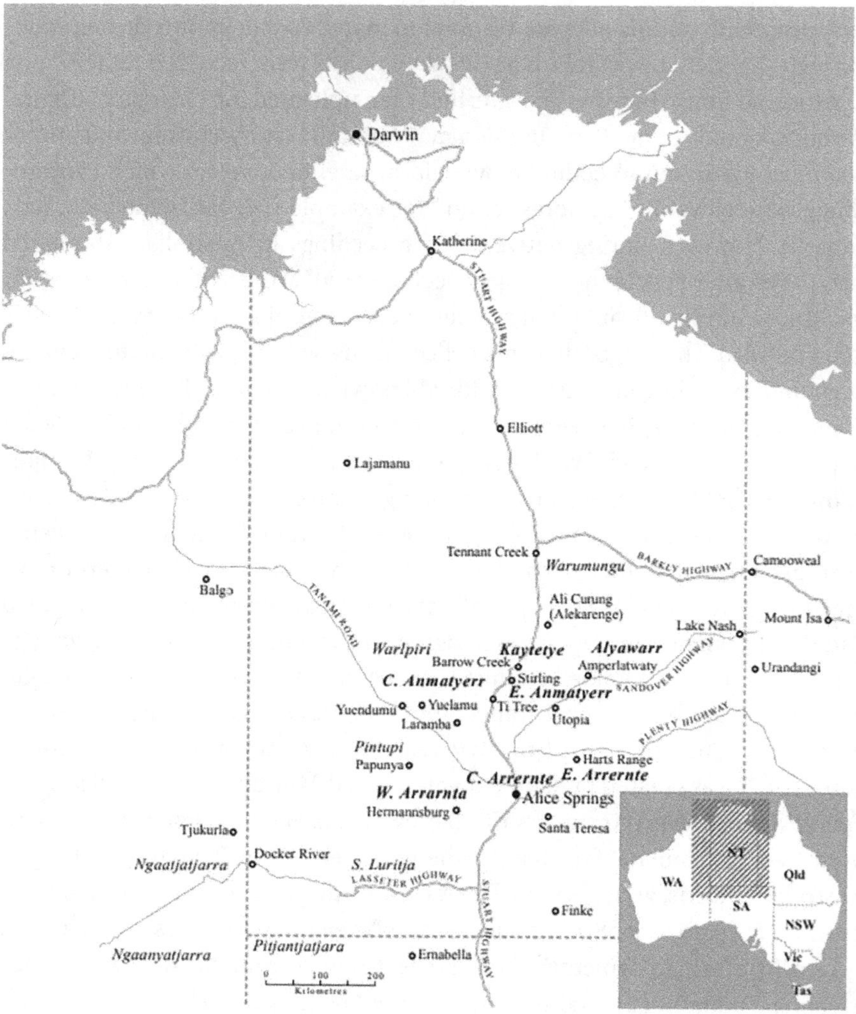

FIGURE 44 Map illustrating the location of Alyawarr, Eastern Anmatyerr, and neighbouring languages

SOURCE: FROM GREEN, 2014, P. 5. COPYRIGHT 2014 BY CAMBRIDGE UNIVERSITY PRESS. REPRINTED WITH PERMISSION

between each other as they can speak or hear Alyawarr as a "second or third language" (Green et al., 2019, p. 1). The neighbouring language groups in Figure 44, like the songlines in the map described above, demonstrate the vast areas where traditional ecological knowledge can be exchanged.

Speaking for the collective needs of First Nations can involve learning "new ways" to map important plant and animal resources (High, 2020, p. 320). Digital mapping, for instance, using "kowori technology", provides a way for outsiders to understand what First Nations such as the Waorani people in Amazonian

Ecuador "know through experiences" (High, 2020, p. 319). Similarly, Martu people in the Australian Western Desert use satellite telemetry to assess the impact of frequent cool burns (small-scale, low-intensity fires) on desert ecosystems (Kanyirninpa Jukurrpa, n.d.-b). The collective needs of Martu are articulated in a short video produced for Kanyirninpa Jukurrpa (n.d.-b), a Martu organisation, that explains how intergenerational knowledge transfer of "fire is more important for Martu people" (Kanyirninpa Jukurrpa, n.d.-b, 01:20). Ranger teams "go out into country to burn the country, to look after it" (02:15), which helps to monitor critically endangered marsupial species such as the greater bilby or *mankarr* and the northern quoll or *wiminyji* (Kanyirninpa Jukurrpa, n.d.-a). Low-intensity cool burns also regenerate bush foods.

> For Martu people who live in this land, it's the songline and the land around this area, it's so special to them. We connect with the land – it's our dreaming, our spirits, our culture. It's my life. It's all Martu people's ancestors, our great grandmother's and grandfather's. What the old people told me – to try and look after our culture, law and the land. So we need to protect Martu culture and the land and the community that lives around it.
>
> CLIFTON GIRIGRBA, PARNNGURR (Kanyirninpa Jukurrpa, n.d.-a, para. 4)

For Alyawarr, research as part of a land claim undertaken in Central Australia (the Alyawarra and Kaititja land claim; see Toohey, 1979) demonstrates "the relationship between songs and land ownership" (Higgins, 2021, p. 727). The relationship is understood as described by Moyle (1983):

> There are people who are said to own songs. Other people must ask their permission to perform those songs. The people who own those songs own the ceremonies where the songs are sung. The texts of songs relate to Creation myths and other stories. These song texts can be shown on a map. The owners of the ceremonies own the places where the songs travelled through. (p. 6)

In Alyawarr, customary law which encodes the relationship of people to land (Green et al., 2019) – for example, "land owners, whose rights derive from their fathers, and the managers, whose rights are inherited from their mothers" – is not easily recognised by Australian common law (Higgins, 2021, p. 727). While songlines and ceremonies associated with these are considered in land claims, it is difficult to recognise land ownership through cultural practices such as song, dance, or art as these are not easily understood according to the terms under common law (Higgins, 2021). Australian common law relies on evidence that is articulated and recognisable as written deeds rather than performed (through song or dance) or painted as symbolised forms of deeds (Higgins, 2021, pp. 727–730).

Records of "Dreaming lines" or "the routes taken by Dreamtime characters in their travels" (Moyle et al., 1986, p. 41, as cited in Stubington, 1987, p. 73) are documented, so "the songs and ceremonies which give sensory perception to the myths may be placed in the perspective of a geographical framework" (Moyle et al., 1986, p. 43, as cited in Stubington, 1987, p. 73). However, more than just important in a land claim,

> ceremonies are very much involved with the passing on of knowledge to those who are entitled to receive it, the rehearsal of knowledge already gained to ensure that it is kept in the correct form, the social organisation of communities and who is entitled to say and do what. (Gray, 1999, p. 6)

Therefore, privileging the voices of Utopia Eastern Anmatyerr and Alyawarr elders helped to maintain the social organisation of communities and to rehearse the knowledge of akatyerr sustainable development needed for the next generations.

2 Lessons for Future Teachers

Adjustments to my pedagogic practice to link curriculum that embedded first languages while on a half-day visit to an important akatyerr harvest site created new "opportunities for students to learn through" (NTBOS, 2017, p. 28). Students conducted a quadrat survey of the vegetation at the akatyerr site and student artefacts captured their understanding through multimodal representations. The representations created an "opportunity to recognize and explore the agency" of students' learning (Pantaleo, 2013, p. 371). Multimodal artefacts (including dialogue) provided many opportunities for students "to represent and express their learning" (Pantaleo, 2013, p. 371). The Utopia Eastern Anmatyerr and Alyawarr adults (assistant teachers and community members) guided what students learned. This collaborative approach privileged and responded to the students' "cultural knowledge" and lived "experience" relating to harvesting akatyerr (Gay, 2013, p. 49). Utopia Eastern Anmatyerr and Alyawarr language and epistemologies sat alongside Western language and epistemologies in measurement curriculum and pedagogic practice by initiating changes that created a space for direct dialogue to inform how this could be achieved.

Ongoing collaboration with adult Utopia Eastern Anmatyerr and Alyawarr participants guided the ways local cultural knowledge relating to akatyerr was contextualised to measurement learning. Contextualising mathematics curriculum relating to akatyerr involved "negotiating in particular places at particular times with particular people present" (Verran, 2013, p. 35) to determine the appropriate ways to merge knowledge. However, the capability to merge, for instance, Yolŋu and Balanda (or Western) worldviews within curriculum requires more than identifying the appropriate integration of knowledge and skills to be enacted in "familiar and highly focused specialist contexts" (Frawley & Fasoli, 2012, p. 316). Time is needed to distil ideas for meaningful measurement teaching and learning opportunities to be created. Additionally, the Balanda may have to challenge their unconscious bias that they "always have cognitive authority" while preparing curriculum that intersects knowledge systems (MacMahon, 2013, p. 23). For instance, consideration may need to be given to the "interaction of linguistic, symbolic and visual systems" when constructing new measurement knowledge (Harper & Parkin, 2019, p. 122). New measurement knowledge may depend on the way movement between these systems is mediated (Harper & Parkin, 2019).

In my study, the use of a real-life context (akatyerr harvest) provided the opportunity to transform measurement curriculum and pedagogy. Students were supported to construct new measurement learning by being provided with ways to navigate the connections between their language, symbolic, and visual representations of measurement attributes (Harper & Parkin, 2019; O'Halloran, 2015). For instance, students could simultaneously (a) engage in substantive conversation about the real-life context; (b) employ reflexive thinking strategies to record visual representations of a measurement attribute; and (c) create symbolic representations of the measurement attribute investigated. The students' flexible use of language, symbolic, and visual representations strengthened their proficiency with proportional reasoning while accessing the intended measurement curriculum (ACARA, 2020a, 2020b).

The study built opportunities for the students to demonstrate their use of proportional reasoning. Students responded to multiplicative situations by considering each subplot within an array or region as a unit of equal size. Simultaneously, students considered the number of subplots covered in vegetation and the total number of subplots (25) within a quadrat surveyed. That is, students responded to the complex real-world context. This created an opportunity for students to shift from simply counting the number of subplots in an array to understanding the vegetation cover as a fraction of the total number of subplots in the array (Siemon et al., 2005). That is, students could deepen their understanding of "the multiplicative relationship between quantities or measures" when comparing the subplots covered or not covered in vegetation (Australian Government Department of Education, 2020, p. 2; see also Behr et al., 1992).

The integration of knowledge and skills from a highly specialised local Indigenous context and a Western context needs to be responsive to "new and changing circumstances" (Frawley & Fasoli, 2012, p. 316). In my study, being responsive included monitoring the presence of threatening species that were likely to impact the sustainable harvest of akatyerr (see Chapter 5, Section 2.3.3 on grasses). Therefore, dialogue with and critical feedback from Utopia Eastern Anmatyerr and Alyawarr participants about the appropriate ways to contextualise knowledge concepts interrelated to akatyerr with measurement learning, combined with my narrative on reorienting my pedagogic approaches to include this dialogue and feedback, have contributed new knowledge on culturally inclusive measurement curriculum and pedagogic practice.

Indigenous cultural content transfer from the appropriate community members was particularly important at the chosen akatyerr patch because teaching and learning experiences may have occurred near areas of cultural significance. Therefore, it was necessary to consult with appropriate elders to obtain permission and guidance to appropriately manage areas used for sustainable akatyerr harvest. Appropriate cultural content considered in this study aligned with the NTILC curriculum developed in consultation with experienced Aboriginal educators and endorsed by the NTBOS (NTDOE, 2017, p. 2).

From the NTILC framework, the pedagogic practices included selecting "different ways of representing [akatyerr and measurement] knowledge" (NTDOE, 2017, p. 47) through reading and representing estimations of the akatyerr growing landscape and vegetation frequency in graphs and charts. Additionally, students listened to "oral stories" about the akatyerr growing site (Natural Environment, Country and Land, and People and Kinship; NTBOS, 2017, pp. 28–35). By listening to adult participants speak in Utopia Eastern Anmatyerr and Alyawarr languages, students were enacting communication and learning strategies that were considered an appropriate achievement standard (Band 2; see Table 1) for the student cohort. That is, middle years (Years 8–9) students who had not studied an Indigenous language formally from early years (Foundation to Year 2) through to middle years (Years 6–9) (NTDOE, 2017) learned the mathematisation of a traditional act (harvesting wild akatyerr) through listening to dialogue in local Indigenous languages. Intergenerational transfer of akatyerr knowledge, an aspiration of the wider community, was communicated through the continued use of the Utopia Eastern Anmatyerr and Alyawarr languages. Students communicated their multimodal representations – photographs, audiovisual recordings, drawings on a grid, and numeric data entered in a table – by listening and responding to oral English and Utopia Eastern Anmatyerr and Alyawarr languages.

3 Mathematisation of Wild Akatyerr Harvest

Wild harvest or "the collection of plants and fruit that occur naturally in a given landscape (as opposed to farmed plants or enrichment planting)" can be the focus of a secondary school measurement unit (Maclean et al., 2019, p. 37). To facilitate the mathematisation of wild akatyerr harvest, I considered a suitable number of quadrats, quadrat points, and transects for the akatyerr investigation for a half-day akatyerr site visit. I developed the data sheets for students to record their observations during their akatyerr quadrat study. The quadrat size (1 m^2) was selected so that it did not "contain too many individual plants ... to count accurately" (Bureau of Land Management [BLM], 1999, p. 12). I built quadrat frames to delineate the quadrat size (100 cm × 100 cm). The frames could be readily carried along the 50-metre transect line to find random quadrat locations (points). Figure 45 illustrates how Student 3 and Student 4 aligned a 1 m^2 quadrat frame at a quadrat point along the transect line to conduct quadrat sampling (see also Arthur, 2020–2021, p. 1).

The transect was laid by using a single 50-metre tape measure, which was held straight and stretched over the area of akatyerr growing land under investigation. The length of transect was chosen based on the "densities and types of vegetation" at the site (BLM, 1999, p. 64). Since akatyerr and additional understory vegetation were sparse, a 50-metre-long transect was used (BLM, 1999). The quadrat sampling involved students' estimations of the "percentage of

FIGURE 45 Aligning the quadrat frame at a quadrat point demonstrated by Student 3 (left). Placing the quadrat frame along transect line demonstrated by Student 4 (right)
PHOTOGRAPHS BY TIM WARE [LEFT] AND STUDENT 3 [RIGHT]

ground surface covered by vegetation" or vegetation cover within the quadrat frame (BLM, 1999, p. 25).

"Close-up" photographs taken by students showed the "amount of ground surface covered by vegetation" (BLM, 1999, p. 31). Students used the quadrat study to investigate ways to promote sustainable akatyerr harvest. This included investigating the distribution of invasive vegetation such as grasses that are highly competitive (e.g., Buffel grass or *Cenchrus ciliaris*; NTG, 2018, p. 10). Additionally, the quadrat study enabled students to investigate threatened vegetation such as *Typhonium*, a unique and undescribed remnant rainforest species at the akatyerr site (NTG, 2018, p. 8). Close-up photography is a skill that students can use in other biological surveys – for instance, of marine environments. Moreover, students can gain qualifications through registered training organisations (RTO) such as Envirotech Education. This RTO trains students in preparing photo quadrats as part of a VET qualification, the Certificate II in Marine Habitat Conservation and Restoration (Envirotech Education, n.d.).

Land management activities (and transferable skills) were embedded in the akatyerr site visit. Adult participants discussed and identified the appropriate location to survey akatyerr vegetation. Utopia Eastern Anmatyerr and Alyawarr assistant teachers and non-Indigenous teaching staff coordinated student involvement in the contextualised measurement learning that occurred at the akatyerr site to ensure cultural obligations in relation to the site were maintained. Adjusting the pedagogic design to include the use of quadrats enabled students to trial methods to survey and record observations of akatyerr plant distribution and yield. Student artefacts were created by four students during the akatyerr site quadrat study. The length in metres along the transect line was recorded as a quadrat point, and the area of akatyerr or other vegetation plant cover was recorded as a percentage. Photographs and drawings of vegetation cover assisted students to think reflexively about what they observed within each quadrat. Visual, numeric, and oral representations of the percentage cover estimated by students were recorded for each quadrat sampled. The visual representations were captured in photographs and drawings on a grid (as an array). The numeric representations were captured as symbolised patterns entered in a table. The oral representations were captured as spoken two-way dialogue between the students and me. Each multimodal representation captured the students' progression in measurement knowledge when contextualised to their akatyerr harvest knowledge interests.

The research instruments developed for the akatyerr investigation included options for students to measure the following attributes: mass, capacity, length, area, perimeter, and volume. However, at the time we visited the akatyerr site the plants we found were not ready for harvest. The mass of dried akatyerr was therefore recorded as 0 g/m^2. The capacity as a handful or bucketful of dried akatyerr was assessed as 0 handful/m^2. It was possible to measure the length and area of akatyerr vegetation and other plants observed at the site. Quantifying the changes in relative abundance of plants found at the site over time was a monitoring strategy that enabled students to trial current land management practices used to care for country (Philippoff & Cox, n.d.). Students asked the community members how to care for country and conducted their own investigation that assessed possible threats to the country (the percentage cover of invasive grasses). The area of quadrat covered in akatyerr or other vegetation was estimated. The volume of rainwater that could be harvested on the akatyerr site was determined from the length and width dimensions of the quadrat frame. The volume for a rectangular prism with a 100 cm × 100 cm base was able to be determined after the akatyerr site visit (see Figure 46). Figure 46 across the following two pages shows a culturally inclusive work sample. The work sample was adapted to my study from a different context provided by ACARA (2014). I provide this work sample to show how you can adapt a typical "above satisfactory" response expected for this achievement standard to include the akatyerr harvest context for solving problems relating to the volume of rain falling on a square prism (ACARA, 2014, pp. 27–28).

Although not part of the study, the measurement attribute of angle could also be evaluated after the akatyerr transect study with the use of satellite imagery such as that shown in Figure 47. The attribute of angle relates very closely to the mathematics topic of geometry (Yumi Deadly Centre, 2014, p. 93). The measurement of angles at the corners of a two-dimensional shape such as a rhombus, combined with the study of trigonometry, makes it possible to calculate other measures such as area once the size of a particular angle is known (Yumi Deadly Centre, 2014, p. 96). The area of the akatyerr site investigated, depicted in Figure 47 as a rhombus, was able to be assessed using the length of one of the sides of the rhombus (as the 50 m transect length) overlaid on an aerial photograph (Figure 47). An exploration of Euclidean geometry was possible using the parallel postulate and geometric reasoning where all sides of a rhombus have the same length (congruent) and opposite sides are parallel, as depicted in the aerial photograph in Figure 47.

The perimeter of a parallelogram can be calculated using the formula $P = 2(a + b)$ or of a rhombus using the formula $P = 4a$ where a or the base is the transect line (if you rotate the figure) and b or the sides are the connecting lines (black lines). The length of the transect line overlaying the aerial map is 2 cm. The 2 cm red line represents the 50-metre transect line used at the akatyerr site. Figure 48 shows a section of the aerial image of the akatyerr harvest site (from Figure 47) with parallel transect lines (red lines) and connecting lines

Year 8 Inclusive Mathematics

Measurement: Rain on the akatyerr site

Year 8 Mathematics achievement standard

The parts of the achievement standard targeted in the assessment task are italicised:

By the end of Year 8, students solve problems relating to the volume of prisms

Summary of task:

Students complete this task in class time after authentic learning experiences at an akatyerr harvest site.

Transect length (m)	Akatyerr growing area (m^2)
1	1
50	50
150	150
300	300

Assume the land is flat (little difference to the rain collected)

Choose the transect length

Calculate the amount of rainwater in litres (L) collected when 1 millimetre (1 mm) of rain falls on the akatyerr growing area chosen.

FIGURE 46 Culturally inclusive Year 8 work sample – solving problems relating to the volume of prisms

Year 8 Inclusive Mathematics **Work sample**

My choice of Akatyerr growing area is:
$1\ m^2$ (1 m transect length)

Rain collected on akatyerr growing area is a square prism

Calculations

Volume of rain $(1 \times 0.001)\ m^3$

1 mm = 0.001 m
$1\ m^3$ holds = 1000L

Volume in litres $(1 \times 0.001 \times 1000)$ L = 1 L

The amount of rainwater collected when 1 mm of rain falls on the akatyerr growing area is: the same as the akatyerr growing area because the 0.001 (height of the square prism) is cancelled out when multiplying by 1000

FIGURE 46 Culturally inclusive Year 8 work sample – solving problems relating to the volume of prisms (*cont.*)

FIGURE 47 Aerial image of akatyerr harvest site with parallel transect lines (red lines)
SOURCE: IMAGE CAPTURED FROM GOOGLE MAPS BY SAMANTHA KENDAL [CENTRAL LAND COUNCIL]

(black lines) – transformed (flipped then rotated) and represented to solve area of rhombus.

Using rotational symmetry and depicting a rhombus with pairs of congruent angles opposite each other allows the akatyerr growing area parallel to a transect line to be measured. Students, future generations of land managers, elders, or school staff can then calculate the land size of the akatyerr growing site. For instance, if you flip and rotate the section of the aerial photograph depicted in Figure 48, you can then apply the trigonometry formula for area of a rhombus as shown in Figure 49:

If the length of one of the sides of the rhombus is known and one of the angles is known, then the trigonometry formula for area of a rhombus can be applied. Since the length of one side of the rhombus is known ($S = 50$), and one of the included angles is known (angle <ABD = 100°), then the area of the rhombus can be calculated (Area = $S^2 \sin B$). That is, Area = $50^2 \times \sin(100°)$. This calculates the area of the rhombus as 2,450 m^2. Students could later estimate the vegetation cover of akatyerr and other plants growing within the broader akatyerr growing site (rhombus area of 2,450 m^2) based on the percentage cover assessed in random sampling within 1 m^2 quadrat frames. Knowing the

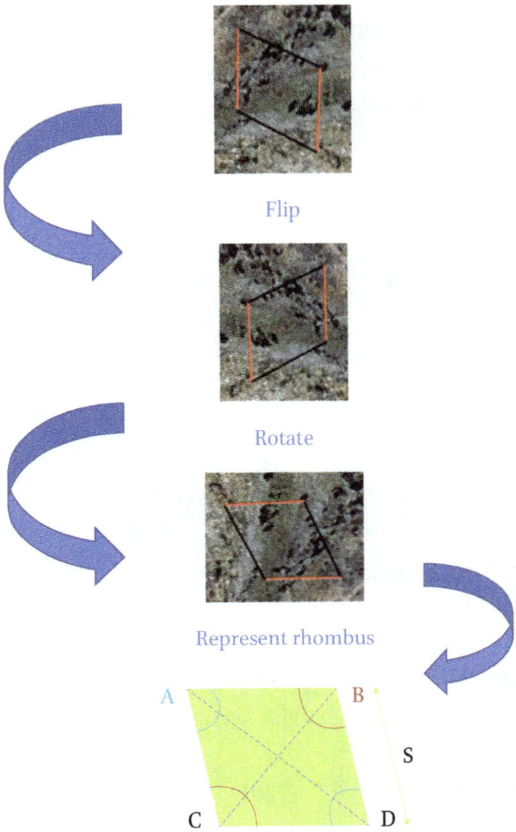

FIGURE 48
A section of the aerial image of akatyerr harvest site transformed and represented to solve area of rhombus (by applying trigonometry formula as Area = $S^2 \sin A$ or Area = $S^2 \sin B$)
SOURCE: IMAGE CAPTURED FROM GOOGLE MAPS BY SAMANTHA KENDAL [CENTRAL LAND COUNCIL]

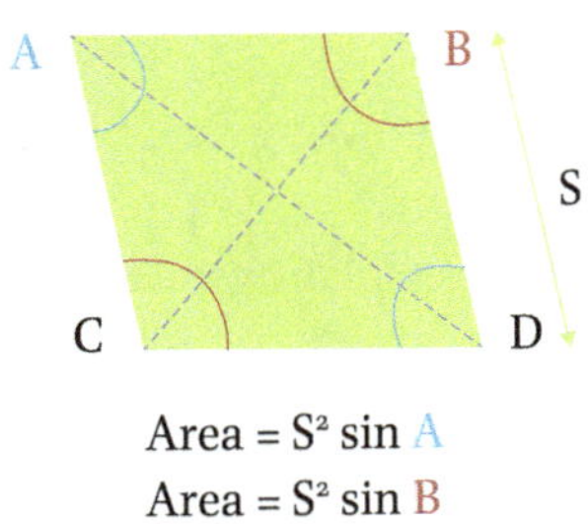

FIGURE 49
Geometric reasoning (congruent angles) and solving area of a rhombus (Area = $S^2 \sin B$). That is, if S (side length) is 50 and angle <ABD is 100° then Area = $S^2 \times \sin(100)$

dimensions of the area defined within the rhombus can help determine where to focus efforts on the protection of seeds and fruit contained within the site – for instance, to target burning efforts and removal of invasive plant species, such as Buffel grass (*Cenchrus ciliaris*). In Central Australia, "Aboriginal knowledge and use of fire to manage landscape" (Smith et al., 2019, p. 294) is particularly important to provide "better conditions" for traditional food (Hunt et al., 2009, p. 10), such as the bustard (*Ardeotis australis*) or emu (*Dromaius novaehollandiae*), which feed on akatyerr fruit. Knowing the land size of the akatyerr growing site is also important to verify the significance of this site for future Indigenous Protected Area applications.

4 Future Questions

For Utopia elders:
- What are the akatyerr harvesting rules?
- How do you care for this country?
- Where are the water places?
- How do we promote bush animals?
- What are the plant names?
- How much does it rain?
- Should you burn this country? How and where?
- What information can be stored on data collection apps?

For Utopia Alyawarr rangers:

Traditionally, Eastern Anmatyerr and Alyawarr people managed their land in Utopia with fire. They lit low-intensity fires or cool burns at regular time intervals.
- What are the benefits of regular cool burns for Utopia Eastern Anmatyerr and Alyawarr people?
- What are the benefits for the desert ecosystem?
- When was this country burnt recently? In the past?
- What plants were growing here before? What plants are growing here now? Is there new plant growth?
- Have animals come back after the burn? Is there fresh plant growth for animals?
- Is there more food to harvest now or in the past?
- Why do you think this is?
- What information can be stored on data collection apps?

For students and school staff in Utopia:
- Is it possible to package and sell dried akatyerr (100 g bags) to tourists through the local Aboriginal corporation?
- How is pricing of akatyerr dependent on the market?
- How do you observe and interpret what is happening on the national market?
- What are the market opportunities for akatyerr sale?

- Where do we go to find out the market research? Websites? Contacts?
- What potential is there for secondary students to be involved in vocational training in conservation and land management, governance, tourism and/or using their first language skills and communicating and translating cultural knowledge to English?
- Can students and school staff develop digital literacies while sharing knowledge about country?
- Can video-recorded footage on bush food collection (during school excursions) and stories shared by elders be stored and managed through different digital media in school and while on country?
- How do you protect Indigenous data sovereignty?
- Can students and school staff be involved in future wild harvest workshops?

For land management authorities/practitioners (such as Rangeland Scientists at the Arid Zone Research Institute):
- What mapping and data services are available for Indigenous rangers?
- What information is collected on land management activities? Fire? Weeds? Feral animals? Water resources? Plant diversity? Animal species? Habitats?
- What are the data-sharing protocols? (see Janke, 2021, p. 303)
- What opportunities are there for innovation and collaboration with Indigenous ranger groups, schools, and community groups in threatened species management and recovery plans?

For teachers involved in decolonising education, Rigney (2019, slide 50) provides the following advice:
- Can all students especially Aboriginal children see a positive sense of their identity reflected in your centre?
- How much prior knowledge and intelligences of the child are your pedagogies capturing?
- How does your teaching establish an environment for them to realise their democratic rights?
- How do you sustain family and elder relationships?
- How do you connect cultural and linguistic repertoire to learning?

5 Summarising Thoughts

Like many teachers with a Western worldview, I struggle against the transmission of cultural norms in teaching practices that view and judge non-Western cultures as inferior against a standard imposed by my own Western lens (Lewis & Aikenhead, 2000). Additionally, I am challenged by the tendency to view Indigenous perspectives as only relevant through the relationship of these views to this imposed standard. This study, however, positioned me to critically reflect on my own story as a teacher. I have come to know and understand culturally inclusive curriculum and pedagogical practices suitable for my student cohort since I listened and responded to students' interests in and lived experiences of harvesting akatyerr.

It has been an honour and a privilege to learn from the Utopia Eastern Anmatyerr and Alyawarr people involved in this study and in the school community where I worked and lived. The following is a reflection provided by Victor Steffensen, a descendant of the Tagalaka people through his mother's country in northern Queensland (Janke, 2021). As an author of the book *Fire Country*, Victor wrote that for the last two remaining Awu-Laya elders it was critical:

> to practise culture and get back onto their country. They wanted to apply their knowledge back onto the land, the fire, the water, looking after the story places. But most of all they wanted their younger ones to learn the language and get back onto country. It was vital because the two men were the last of the Awu-Laya Elders who knew the traditional knowledge and stories of that country. They wanted the young ones to inherit the knowledge and take over their roles as leading Elders. (Steffensen, 2020, p. 11)

We must listen to and respect the wisdom and aspirations of elders. While advocating for pedagogies which honour and integrate worldviews – as educators, we can incorporate the stories shared by *all* global citizens.

References

Arthur, M. (2020–2021). *Transects and quadrats rock platform 1–2 lessons: Measuring distribution and abundance* [Biology II course notes]. Studocu, University of New England. Retrieved January 29, 2023, from https://www.studocu.com/en-au/document/university-of-new-england-australia/biology-ii/transects-and-quadrats-rock-platform-1-2-lessons/13086905

Australian Curriculum, Assessment and Reporting Authority. (2014). *Australian curriculum: Work sample portfolio – Mathematics Year 8: Above satisfactory*. Retrieved January 11, 2023, from https://docs.acara.edu.au/curriculum/worksamples/Year_8_Mathematics_Portfolio_Above.pdf

Australian Curriculum, Assessment and Reporting Authority. (2020a). *Australian curriculum: Learning continuum of numeracy: Using fractions, decimals, percentages, ratios and rates (Level 5)*. Retrieved January 11, 2023, from https://bit.ly/3GzjB4i

Australian Curriculum, Assessment and Reporting Authority. (2020b). *Australian curriculum: Mathematics (Version 8.4) – Year 8 level description*. Retrieved January 11, 2023, from https://bit.ly/3GB8zM7

Australian Government Department of Education. (2020). *Re(Solve) maths by inquiry – What you need to know: Proportional reasoning*. Retrieved January 10, 2023, from https://www.resolve.edu.au/sites/default/files/downloads/Proportional%20Reasoning%20-%20Learning%20Progression%20and%20WYNTK.pdf

Behr, M., Harel, G., Post, T., & Lesh, R. (1992). Rational number, ratio and proportion. In D. Grouws (Ed.), *Handbook of research in mathematics teaching and learning* (pp. 296–333). MacMillan.

Bureau of Land Management. (1999). *Sampling vegetation attributes* (Interagency Technical Reference 1734-4). Bureau of Land Management National Applied Resource Sciences Center. Retrieved January 10, 2023, from https://www.ntc.blm.gov/krc/uploads/296/TR%201734-04%20Sampling%20Veg%20Attributes%20%20(1996,%20revised%201999).pdf

Cornell, S., & Jorgensen, M. (2019). What are the limits of social inclusion? Indigenous peoples and Indigenous governance in Canada and the United States. *American Review of Canadian Studies, 49*(2), 283–300. https://doi.org/10.1080/02722011.2019.1613790

Digital Democracy. (n.d.). *Waorani case study*. Retrieved January 7, 2023, from https://www.digital-democracy.org/case-study/waorani-case-study

Envirotech Education. (n.d.). *Marine, environment & sustainability: 10749NAT – Certificate II in marine habitat conservation and restoration*. Retrieved January 11, 2023, from https://envirotech.edu.au/en/configuration/courses/10749nat-certificate-ii-in-marine-habitat-conservation-and-restoration/

First Nations Education Steering Committee & First Nations Schools Association. (2020). *Math First Peoples teacher resource guide: Elementary & secondary*. Retrieved January 11, 2023, from http://www.fnesc.ca/math-first-peoples/

Frawley, J., & Fasoli, L. (2012). Working together: Intercultural leadership capabilities for both-ways education. *School Leadership & Management, 32*(4), 309–320. https://doi.org/10.1080/13632434.2012.708328

Gay, G. (2013). Teaching to and through cultural diversity. *Curriculum Inquiry, 43*(1), 48–70. https://doi.org/10.1111/curi.12002

Gray, P. (1999). Aboriginal and native title issues. *Australian Law Librarian, 7*(1), 5–13. Retrieved January 11, 2023, from https://search.informit.org/doi/abs/10.3316/agis_archive.19993146

Green, J. (2014). *Drawn from the ground: Sound, sign and inscription in Central Australian sand stories.* Cambridge University Press. https://doi.org/10.1017/CBO9781139237109

Green, J., Blackman, D., & Moore, D. (2019). *Alyawarr to English Dictionary* (2nd ed.). IAD Press.

Harper, H., & Parkin, B. (2019). Talking and writing to develop mathematical meanings in a remote Indigenous context. In J. Rennie & H. Harper (Eds.), *Literacy education and Indigenous Australians* (pp. 109–125). https://doi.org/10.1007/978-981-13-8629-9_7

Higgins, N. (2021). Songlines and land claims; space and place. *International Journal for the Semiotics of Law, 34*, 723–741. https://doi.org/10.1007/s11196-020-09748-z

High, C. (2020). "Our land is not for sale!" Contesting oil and translating environmental politics in Amazonian Ecuador. *The Journal of Latin American and Caribbean Anthropology, 25*(2), 301–323. https://doi.org/10.1111/jlca.12494

Hunt, J., Altman, J., & May, K. (2009). *Social benefits of Aboriginal engagement in natural resource management* (Working Paper No. 60/2009). Centre for Aboriginal Economic Policy Research, Australian National University. Retrieved January 25, 2023, from https://caepr.cass.anu.edu.au/research/publications/social-benefits-aboriginal-engagement-natural-resource-management

Janke, T. (2021). *True tracks: Respecting Indigenous knowledge and culture.* NewSouth Publishing.

Kanyirninpa Jukurrpa. (n.d.-a). *Martu country.* Retrieved January 3, 2023, from https://www.kj.org.au/martu-country

Kanyirninpa Jukurrpa. (n.d.-b). *Waru (fire) program* [Video]. Retrieved January 3, 2023, from https://www.kj.org.au/waru-fire-program

Lewis, B., & Aikenhead, G. (2000). Introduction: Shifting perspectives from universalism to cross-culturalism. *Science Education*, 85, 3–5.

Maclean, K., Woodward, E., Jarvis, D., Rowland, D., Rist, P., Turpin, G., Martin, P., & Glover, R. (2019). *A strategic sector development and research priority framework for the traditional owner-led bush products sector in northern Australia.* CSIRO. https://doi.org/10.25919/5d72acc375ac7

MacMahon, K. (2013). The promise of milmarra. *Learning Communities: International Journal of Learning in Social Contexts, 12*(April), 18–23. Retrieved January 10, 2023, from https://digitalcollections.cdu.edu.au/nodes/view/4735

Morris, C. (2007). *A dialogical encounter with an Indigenous jurisprudence* [PhD thesis, Griffith University]. Griffith Theses. https://doi.org/10.25904/1912/1555

Moyle, R. (1983). Songs, ceremonies and sites: The Agharringa case. In N. Peterson & M. Langton (Eds.), *Aborigines, land and land rights* (p. 6). Australian Institute of Aboriginal Studies. Retrieved January 7, 2023, from https://catalogue.nla.gov.au/Record/2161285

Northern Territory Board of Studies. (2017). *Keeping Indigenous languages and culture strong: A plan for the teaching and learning of Indigenous languages and cultures in Northern Territory Schools*. Northern Territory Government. Retrieved January 11, 2023, from https://education.nt.gov.au/__data/assets/pdf_file/0012/413202/Policy_Keeping-Indigenous-Languages-and-Cultures-Strong-Document_web_updated.pdf

Northern Territory Department of Education. (2017). *NT Indigenous languages and cultures: First languages maintenance (L1M)*. Northern Territory Government. Retrieved January 11, 2023, from https://education.nt.gov.au/__data/assets/pdf_file/0020/715232/ntilc-l1m-nov18-03.pdf

Northern Territory Government. (2018). *Angarapa ALT [Aboriginal Land Trust] NT NRM [Northern Territory Natural Resource Management] report*. Retrieved July 22, 2019, from NT Infonet: http://www.infonet.org.au

O'Halloran, K. (2015). The language of learning mathematics: A multimodal perspective. *Journal of Mathematical Behavior*, *40*(Part A), 63–74. https://doi.org/10.1016/j.jmathb.2014.09.002

Pantaleo, S. (2013). Matters of design and visual literacy: One middle years student's multimodal artifact. *Journal of Research in Childhood Education*, *27*(3), 351–376. https://doi.org/10.1080/02568543.2013.796334

Philippoff, P., & Cox, E. (n.d.). *Measuring abundance: Transects and quadrats. Concepts: estimation, percentage, error and zonation*. University of Hawaii. Retrieved January 10, 2023, from https://www.hawaii.edu/gk-12/opihi/classroom/measuring.pdf

Rigney, L. I. (2019, September 25–28). *Early years practice using Aboriginal learner cultural strengths? Culturally responsive pedagogies in dialogue with Reggio Emilia* [Australian keynote address PowerPoint slides]. Early Childhood Australia National Conference 2019, Hobart, Tasmania, Australia. Retrieved December 19, 2022, from https://www.ecaconference.com.au/2019/presenter-papers-2019/

Siemon, D., Breed, M., & Virgona, J. (2005). From additive to multiplicative thinking: The big challenge of the middle years. In J. Mousley, L Bragg, & C. Campbell (Eds.), *Mathematics – celebrating achievement: Proceedings of the 42nd conference of the Mathematical Association of Victoria*. MAV.

Smith, A., McNiven, I. J., Rose, D., Brown, S., Johnston, C., & Crocker, S. (2019). Indigenous knowledge and resource management as world heritage values: Budj Bim cultural landscape, Australia. *Archaeologies*, *15*, 285–313. https://doi.org/10.1007/s11759-019-09368-5

Steffensen, V. (2020). *Fire country: How Indigenous fire management could help save Australia*. Hardie Grant Travel.

Stubington, J. (1987). Alyawarra music: Songs and society in a central Australian community. *Musicology Australia*, *10*(1), 73–74. https://doi.org/10.1080/08145857.1987.10415182

Sutton, P. (1998). Aboriginal maps and plans. In D. Woodward & G. Malcolm Lewis (Eds.), *The history of cartography: Vol. 2, Book 3. Cartography in the traditional African, American, Arctic, Australian and Pacific societies* (pp. 387–416). Retrieved January 7, 2023, from https://press.uchicago.edu/books/HOC/HOC_V2_B3/HOC_VOLUME2_Book3_chapter10.pdf

Toohey, J. (1979). *Land claim by Alyawarra and Kaititja: Report by the Aboriginal Land Commissioner to the Minister for Aboriginal Affairs*. Commonwealth of Australia. Retrieved January 11, 2023, from https://www.dss.gov.au/sites/default/files/documents/05_2012/03.pdf

Verran, H. (2013). The dead as participants: Challenged by the Yolŋu Aboriginal child learner at Gangan. *Learning Communities: International Journal of Learning in Social Contexts, 12*(April), 35–39. Retrieved January 10, 2023, from https://researchers.cdu.edu.au/en/publications/the-dead-as-participants-challenged-by-the-yolngu-aboriginal-chil

Walsh, F. J., Dobson, P. V., & Douglas, J. C. (2013). Anpernirrentye: A framework for enhanced application of Indigenous ecological knowledge in natural resource management. *Ecology & Society, 18*(3), Article 18. https://doi.org/10.5751/ES-05501-180318

YuMi Deadly Centre. (2014). *YuMi Deadly Maths AIM overview: Objectives, acceleration framework, pedagogy, culture and school change, implementation options*. Queensland University of Technology. Retrieved December 9, 2022, from https://research.qut.edu.au/ydc/wp-content/uploads/sites/181/2019/12/AIM-Overview-2014.pdf

Glossary

Quadrat frame a frame, usually square, used in ecology to isolate a standard unit of area for the study of the distribution of plants over a large area.

Quadrat point when a quadrat is placed on the ground at a quadrat point along the transect line the plants present within the frame are identified and their abundance (% cover) recorded.

The study refers the original research conducted to fulfill a Master of Philosophy (Education) degree at QUT in 2021.

Transect lines are a long narrow sampling area, so can extend over a long portion of the habitat. Line transects are used to record the changing pattern of plants and animals within a habitat.

Utopia Eastern Anmatyerr and Alyawarr is used in this study to specifically identify the language groups of First Peoples of the central Australian desert region where the study was situated.

Western perspective the tendency for Anglo Celtic Australians, for instance, to focus more on compartmentalized knowledge than understanding the bigger, related picture.

Index

Aboriginal and Torres Strait Islander Mathematics Alliance (ATSIMA) 1, 142
Acacia georginae 10
agreement-making 68, 89, 90
aherr 5
akatyerr 5, 7–9, 12, 15–17, 19, 20, 30, 32–37, 40, 42–50, 52, 54, 63, 64, 67, 71–77, 80, 81, 90–106, 110–121, 123–128, 132–134, 136–141, 143, 147–165, 170, 171, 174–184, 186
Alice Springs (Mparntwe) Education Declaration 20, 30, 71
Alyawarr 2, 4, 5, 7–9, 12, 15, 19, 34–37, 39, 40, 42, 45–48, 50, 63, 74–76, 78, 89–98, 101, 114, 115, 133–136, 138, 139, 147, 148, 154, 161, 165, 171, 172, 174–178, 184, 186
Alyawarra and Kaititja Land Claim 1979 173, 174
amern mwerr 5
anaty 5
Anmatyerr 2, 4, 5, 7–9, 12, 15, 19, 34–37, 39, 40, 42, 45–48, 50, 63, 74–76, 78, 89–98, 101, 114, 115, 133–136, 138, 139, 147, 148, 154, 161, 165, 171, 172, 174–178, 184, 196
Apmer Akely-akely 4, 65, 67–69
Apmikw 68
Arid Zone Research Institute 185
Arlewatyerr 5
Arnhem Land 101
Arrernte 7, 63, 161
arreth 68
artworks 10, 70–72, 80, 94, 140, 158
arwngerrp 5
Australian Curriculum, Assessment and Reporting Authority (ACARA) 13, 52, 103
Australian Curriculum: Mathematics 16, 157, 162
Australian Government 6, 7, 30, 31, 38, 165
Australian Institute of Teaching and School Leadership (AITSL) 51
Australian Native Food and Botanicals (ANFAB) 67, 68
Awu-Laya Elders 186

Balanda (freshwater people) 175
Bandaiyan 171
Barunga Agreement 89, 90
"Big Business" unit 64, 65, 112
"Big Day Out" 1
bloodwood tree 5, 6
"both-ways" education 2 12, 161
Brown, Rayleen 7, 63, 112, 160, 161
bush balm business 68–70
bush foods industry 67
Buffel grass (*Cenchrus ciliaris*) 9, 73, 178, 183
Bureau of Land Management (BLM) 177, 178
bustard (*Ardeotis australis*) 171, 183

Canada, Australia, New Zealand, and United States (CANZUS) 8
Central Australia Youth Link Up Services (CAYLUS) 2
Central Desert region 1, 7, 8, 63, 140
Central Land Council (CLC) 2
Certificate II in Marine Habitat Conservation and Restoration 178
Close-up photographs 178
closing the gap (when referring to the strategy) 1, 7
co-design 19, 73, 147
Coranderrk 1
country 1, 2, 5–7, 14, 15, 17, 36, 50, 66, 67, 70, 74, 75, 80, 91–93, 95, 100, 101, 103, 105, 106, 110, 115, 126, 128, 136, 150, 151, 170, 171, 173, 176, 179, 184–186
creation time 14, 15, 50, 136, 170
cultural respect framework 33, 34, 47, 48, 138

Department of Primary Industries and Resources 67
Digby, Donna (bushfoods industry development officer) 67
direct instruction 18, 78, 79

Eastern Anmatyerr and Alyawarr people 2, 4, 5, 7–9, 12, 15, 19, 34–37, 39, 40, 42, 45–48, 50, 63, 74–76, 78, 89–98, 101, 114, 115, 133–136, 138, 139, 147, 148, 154, 161, 165, 171, 172, 174–178, 184, 196
Employment Pathways unit 11, 64–69, 79, 112
emu (*Dromaius novaehollandiae*) 171, 183
Eremophila dalyana 9

ethno-ecologist 67, 80
Euclidean geometry 179

Finlay, Summer May 90, 95
first languages 17, 41, 51, 65, 74, 76, 78, 95, 106, 139, 147, 174, 185
First Nations, First Peoples 29, 30, 47, 54, 65, 68, 69, 89, 90, 95, 133–136, 148, 170, 172

Ga<u>n</u>ma 15, 51
geometric reasoning 106, 179, 183
Georgina gidgee (*Acacia georginae*) 9
Girigrba, Clifton 173
Goompi pedagogical cycle 48, 74, 106, 110–115, 118, 121, 126, 127, 148, 152, 154, 157, 159
Grandplover River (pseudonym) 9, 36

Homeland Learning Centres 6
Hyllus Maris 1

ilpengk (*Eremophila datyana*) 66, 68
Indigenist research principles/framework 34, 38, 44, 47, 54, 94, 98
Indigenous cultural and ecological knowledge (ICEK) 53, 67, 113
Indigenous cultural and intellectual property (ICIP) 34, 50
Indigenous governance awards 4, 69
Indigenous languages and cultures curriculum 5, 13, 16, 75, 76, 128
informed-consent 36, 37, 45, 50, 97, 138, 139
intergenerational 5, 7, 10, 15, 16, 46, 49, 52, 63, 92, 97, 98, 105, 143, 164, 173, 176
irreyakwerr 5

Janke, Terri 45, 68

Kanyirninpa Jukurrpa 173
Kaytetye 171
Kimberley 171
Kngwarreye, Emily Kame 71
kowori technology 172
Kulin Nation 1

Makarrata 89, 98
Mankarr 173
Martu people 173
Maths and Indigenous Learners Closing the Gap: Make It Count project 1
Matthews, Chris 1, 110, 111, 142, 157
member checking 97, 143
Merne Altyerr-ipenhe Reference Group 14, 15, 49, 50, 63, 67, 92, 93, 136, 153
Minjerribah 110
Mowaljarlai, David 171
multimodal 143, 174, 176, 178

narrative inquiry 132–135, 141, 143
National Assessment Program – Literacy and Numeracy (NAPLAN) 52
Native Title Act 1993 36
Nelson, Esau 4, 69
Ngangiwumirr 7, 63, 161
Ngarinyin people 171
Northern Territory Board of Studies (NTBOS) 12
Northern Territory Department of Education (NTDOE) 5, 16, 65, 142
Northern Territory Government (NTG) 12, 13, 65, 66, 90
Northern Territory Indigenous Languages and Cultures (NTILC) 16, 17, 30, 40, 41, 176
ngkwarl 5, 6
NVivo software 96
nyerralth 19, 41

Organisation for Economic Co-operation and Development (OECD) 52

Parnngurr 173
Patents Act 1990 (Cth) 68
Peeler AM, Dr Lois 1
perpetual pastoral leases 36
Perso, Thelma 78
"Pop-up Shop" 68–71
problem-solving 32, 65, 76, 164
Prostanthera lasianthos 1

quadrat point 114–127, 143, 152–155, 177, 178
quadrat study 47, 74, 114, 120, 143, 152, 155, 159, 177, 178
Quandamooka 1, 110
Queensland University of Technology (QUT) 134, 139, 141

Rangeland scientists 185
registered training organization 178
rock hole 2

sandplains 1
Science, Technology, Engineering, Mathematics (STEM) 77, 141–143, 162
self-efficacy 3, 12, 14, 79, 102
semi-permanent 9, 126
Solanum centrale (akatyerr) 8
songlines 173
Steffensen, Victor 186
Stronger Communities for Children program 4
subplot 117, 123, 175
sub-strands 16, 75, 105, 106
sugarbag 5, 6

Tagalaka people 186
Teaching English to speakers of other languages (TESOL) 142
traditional owners 9, 67, 73, 96, 97, 103, 137
Treaty Barunga Agreement) 89, 90
two-way dialogue 45–47, 95, 98–101, 103, 120, 148, 158, 163, 178
Typhonium (pretty flower) 72, 73, 76, 117, 120–124, 178

United Nations Declaration on the Rights of Indigenous Peoples (UNDRIP) 19, 90
Utopia Desert Art 70, 71
Utopia Eastern Anmatyerr and Alyawarr 2, 8, 9, 12, 15, 19, 34–37, 39, 40, 42, 45–48, 50, 74–76, 78, 89–98, 101, 114, 115, 133–136, 138, 139, 147, 148, 154, 161, 165, 174–178, 184, 186
Utopia Ranger Group 5

video-recorded dialogue 37
vocational education and training (VET) 165, 178

Walsh, Fiona 67, 72, 80
Wanta Aboriginal Corporation 2, 42
Waorani (Ecuador) 170–172
Warlpiri 171
Western (society) 4, 12, 15, 20, 36, 40, 42, 45–48, 51, 54, 95, 102, 114, 140, 173, 174, 176, 185, 186
Western desert 173
white fella 17
wild-harvest activities or industry 63
wiminyji 173
Woiwurrung 1
Worawa Aboriginal College 1
Wunnan 171
Wurundjeri 1

yarning circle 16, 20, 33, 37, 40–42, 45–49, 81, 90–96, 98–103, 105, 106, 114–116, 118, 134, 137–139, 141, 143, 147, 152–154, 156, 158, 160
Yarra Ranges 1
Yolŋu 15, 51, 78, 101, 175
Yorta Yorta 1, 90, 95

www.ingramcontent.com/pod-product-compliance
Lightning Source LLC
Chambersburg PA
CBHW071444230926
51846CB00001B/337
9789004682733